# Kenaf Fibers
and Composites

# Kenaf Fibers
# and Composites

Edited by
S.M. Sapuan
J. Sahari
M.R. Ishak
M.L. Sanyang

CRC Press is an imprint of the
Taylor & Francis Group, an **informa** business

CRC Press
Taylor & Francis Group
6000 Broken Sound Parkway NW, Suite 300
Boca Raton, FL 33487-2742

© 2018 by Taylor & Francis Group, LLC
CRC Press is an imprint of Taylor & Francis Group, an Informa business

No claim to original U.S. Government works

Printed on acid-free paper

International Standard Book Number-13: 978-1-4987-5342-5 (Hardback)
International Standard Book Number-13: 978-1-351-05094-4 (eBook)

This book contains information obtained from authentic and highly regarded sources. Reasonable efforts have been made to publish reliable data and information, but the author and publisher cannot assume responsibility for the validity of all materials or the consequences of their use. The authors and publishers have attempted to trace the copyright holders of all material reproduced in this publication and apologize to copyright holders if permission to publish in this form has not been obtained. If any copyright material has not been acknowledged please write and let us know so we may rectify in any future reprint.

Except as permitted under U.S. Copyright Law, no part of this book may be reprinted, reproduced, transmitted, or utilized in any form by any electronic, mechanical, or other means, now known or hereafter invented, including photocopying, microfilming, and recording, or in any information storage or retrieval system, without written permission from the publishers.

For permission to photocopy or use material electronically from this work, please access www.copyright.com (http://www.copyright.com/) or contact the Copyright Clearance Center, Inc. (CCC), 222 Rosewood Drive, Danvers, MA 01923, 978-750-8400. CCC is a not-for-profit organization that provides licenses and registration for a variety of users. For organizations that have been granted a photocopy license by the CCC, a separate system of payment has been arranged.

**Trademark Notice:** Product or corporate names may be trademarks or registered trademarks, and are used only for identification and explanation without intent to infringe.

**Visit the Taylor & Francis Web site at**
http://www.taylorandfrancis.com

**and the CRC Press Web site at**
http://www.crcpress.com

# Contents

Editors ..................................................................................................................vii
Contributors ..........................................................................................................ix

Chapter 1　Natural Fiber Composites: Challenges and Opportunities ...................1
　　　　　　Faris M. Al-Oqla and S.M. Sapuan

Chapter 2　Kenaf Fiber: Structure and Properties ...............................................23
　　　　　　A.H. Juliana, H.A. Aisyah, M.T. Paridah, C.C.Y. Adrian,
　　　　　　and S.H. Lee

Chapter 3　Adhesion Characteristics of Kenaf Fibers .........................................37
　　　　　　M.T. Paridah and A.H. Juliana

Chapter 4　Kenaf Fiber-Reinforced Thermoplastic Composites ..........................61
　　　　　　A. Khalina and N. Mohd Nurazzi

Chapter 5　Effect of Silica Aerogel on Polypropylene Reinforced
　　　　　　with Kenaf Core Fiber for Interior Automotive Components ...........81
　　　　　　A.S. Harmaen, M.T. Paridah, M. Jawaid, A.M. Fariz,
　　　　　　and B. Asmawi

Chapter 6　Impact of Silane Treatment on the Properties of Kenaf Fiber
　　　　　　Unsaturated Polyester Composites .....................................................93
　　　　　　Md. Rezaur Rahman, Sinin Hamdan, and Rubiyah bt Hj Baini

Chapter 7　Effects of Material Types on the Failure Modes Crashworthiness
　　　　　　Parameters of Kenaf Composite Hexagonal Tubes ..........................113
　　　　　　M.F.M. Alkbir, S.M. Sapuan, A.A. Nuraini, and M.R. Ishak

Chapter 8　Eco-Friendly Kenaf Hybrid Materials .............................................129
　　　　　　S. Norshahida and H. Ismail

Chapter 9　Ballistic Properties of Hybrid Kenaf Composites ............................145
　　　　　　R. Yahaya, S.M. Sapuan, M.R. Ishak, Z. Leman, and M. Jawaid

**Chapter 10** Cellulose-Based Composites from Kenaf Fibers ............................ 169
  *J. Sahari, M.A. Maleque, and M.L. Sanyang*

**Chapter 11** Development and Characterization of Kenaf Nanocomposites ....... 185
  *J. Sahari, M.A. Maleque, and M.L. Sanyang*

**Chapter 12** Concurrent Design of Kenaf Composite Products ......................... 205
  *M.R. Mansor and S.M. Sapuan*

**Index** ............................................................................................................. 227

# Editors

**S.M. Sapuan** is head of the Laboratory of Biocomposite Technology, Universiti Putra Malaysia. He earned his BEng degree in Mechanical Engineering from the University of Newcastle, Australia in 1990, MSc from Loughborough University, UK in 1994 and PhD from De Montfort University, UK in 1998. His research interests include natural fiber composites, materials selection and concurrent engineering. To date, Dr. Sapuan has published more than 450 journal and 450 conference papers, 12 books, 5 edited books and 50 book chapters. He has received numerous awards and honors, among others, ISESCO Science Prize in Technology, 2008; Plastic and Rubber Institute, Malaysia Fellowship Award; Kuala Lumpur Rotary Research Gold Medal Award; Alumni Award, University of Newcastle, NSW, Australia; Khwarizmi International Award (KIA); a Leadership Award from Society of Automotive Engineers International (SAE); SAE Fellow Grade of membership and 2015/2016 grantee of SEARCA Regional Professorial Chair.

**J. Sahari** was a senior lecturer at the Department of Industrial Chemistry, Faculty of Science and Natural Resources, University Malaysia Sabah, Kota Kinabalu, Sabah, Malaysia since 2014. He was appointed Honorary Research Assistant in University of Liverpool, UK in 2012. He earned his Bachelor of Science degree in Industrial Chemistry from Universiti Putra Malaysia, Master of Science in Advanced Material Engineering from Universiti Putra Malaysia in 2011 and PhD also in Advanced Material Engineering from Universiti Putra Malaysia in 2013. Dr. Sahari won the Best Scientific Paper and Oral Presentation of UPM-UniKL MICET Symposium on Polymeric Materials and the Third Postgraduate Seminar on Natural Fibre Reinforced Polymer Composites 2012.

**M.R. Ishak** is a senior lecturer in the Department of Aerospace Engineering, Universiti Putra Malaysia since 2013. He earned his Bachelor's degree in Manufacturing Engineering from Universiti Teknikal Malaysia Melaka in 2007, and Master of Science in Materials and Design Engineering in 2009 and PhD in Materials Engineering in 2012 from Universiti Putra Malaysia. His research interests include natural fiber, biopolymer and biocomposite materials, plastic technology, polymer composite design and testing, composite modifications and properties enhancements, vacuum resin impregnation and manufacturing process. Dr. Ishak registered as a Graduate Engineer with Board of Engineers Malaysia, Graduate Member of Institute Engineers of Malaysia, Member of Malaysian Society of Structural Health Monitoring and Member of Plastics and Rubber Institute of Malaysia. To date, he has 22 journal and 23 conference papers to his credit. He is currently heading a research project titled "Preparation and Properties of Kenaf Santoprene Filled Composites" worth RM 47,000. Dr. Ishak has won the Best Student Poster Presenter in International Conference on Innovation in Polymer Science and Technology (IPST 2011)

International 2011 and has won a silver medal in Exhibition of Invention, Research and Innovation (PRPI 2010) Universiti Putra Malaysia.

**M.L. Sanyang** is currently an Environmental Specialist at the Gambia National Petroleum Corporation (GNPC, Gambia) as well as a part-time lecturer in the Agriculture and Environmental Science School at the University of The Gambia (UTG). He recently completed his post-doctoral fellowship under the Laboratory of Biocomposite Technology at Universiti Putra Malaysia. He earned his Bachelor's degree (BSc) in Material and Mineral Resources Engineering (Petroleum Engineering program) at National Taipei University of Technology (Taiwan), Master's degree (MSc) in Environmental Engineering and PhD in Green Engineering at Universiti Putra Malaysia (UPM). Dr. Sanyang received the "Best PhD Student Award" for his outstanding research, publications and PhD completion on time. His main research interests are: (1) Green Engineering (developing environmentally friendly materials and technology), (2) Environmental Engineering (water and wastewater treatment; utilization of agricultural wastes), and (3) Environmental Management. Dr. Sanyang has published more than 15 journal articles in international journals, authored or co-authored more than 8 book chapters and 2 edited books. He also served as a reviewer for many international journals such as *Polymer Composite*, *BioResources*, *Polymers (MPDI)*, *Polymer Bulletin*, *International Journal of Polymer Science*, *Current Analytical Chemistry*, and others.

# Contributors

**C.C.Y. Adrian**
The World Wide Fund for Nature (WWF), Malaysia
Petaling Jaya, Selangor, Malaysia

**H.A. Aisyah**
Institute of Tropical Forestry and Forest Products
Universiti Putra Malaysia
Serdang, Selangor, Malaysia

**M.F.M. Alkbir**
Department of Mechanical and Manufacturing Engineering
Universiti Putra Malaysia
Serdang, Selangor, Malaysia

**Faris M. Al-Oqla**
Department of Mechanical Engineering
Faculty of Engineering
The Hashemite University
Zarqa, Jordan

**B. Asmawi**
Maero Tech Sdn. Bhd.
Nilai, Negeri Sembilan, Malaysia

**Rubiyah bt Hj Baini**
Faculty of Engineering
Department of Chemical Engineering
Universiti Malaysia Sarawak
Kota Samarahan, Sarawak, Malaysia

**A.M. Fariz**
Faculty of Forestry
Universiti Putra Malaysia
Serdang, Selangor, Malaysia

**Sinin Hamdan**
Department of Mechanical and Manufacturing Engineering
Universiti Malaysia Sarawak
Kota Samarahan, Sarawak, Malaysia

**A.S. Harmaen**
Institute of Tropical Forestry and Forest Products
Universiti Putra Malaysia
Serdang, Selangor, Malaysia

**M.R. Ishak**
Aerospace Manufacturing Research Centre (AMRC)
Faculty of Engineering
and
Department of Aerospace Engineering
Universiti Putra Malaysia
and
Institute of Tropical Forestry and Forest Products
Universiti Putra Malaysia
Serdang, Selangor, Malaysia

**H. Ismail**
School of Materials and Mineral Resources Engineering
USM Engineering Campus
Universiti Sains Malaysia
Nibong Tebal, Penang, Malaysia

**M. Jawaid**
Laboratory of Biocomposite Technology
Institute of Tropical Forestry and Forest Products
Universiti Putra Malaysia
Serdang, Selangor, Malaysia

and

Department of Chemical Engineering
College of Engineering
King Saud University
Riyadh, Saudi Arabia

**A.H. Juliana**
Institute of Tropical Forestry and Forest Products
Universiti Putra Malaysia
Serdang, Selangor, Malaysia

and

Department of Furniture Design and Manufacturing
Faculty of Technology Management and Business
Universiti Tun Hussein Onn Malaysia
Parit Raja, Johor, Malaysia

**A. Khalina**
Faculty of Engineering
Universiti Putra Malaysia
Serdang, Selangor, Malaysia

**S.H. Lee**
Institute of Tropical Foresty and Forest Products
Universiti Putra Malaysia
Serdang, Selangor, Malaysia

**Z. Leman**
Department of Mechanical and Manufacturing Engineering
Universiti Putra Malaysia
Serdang, Selangor, Malaysia

**M.A. Maleque**
Islamic International University of Malaysia
Kuala Lumpur, Malaysia

**M.R. Mansor**
Faculty of Mechanical Engineering
Universiti Teknikal Malaysia Melaka
Hang Tuah Jaya, Melaka, Malaysia

**S. Norshahida**
Department of Manufacturing and Materials Engineering
Faculty of Engineering
International Islamic University Malaysia
Kuala Lumpur, Malaysia

**A.A. Nuraini**
Department of Mechanical and Manufacturing Engineering
Universiti Putra Malaysia
Serdang, Selangor, Malaysia

**N. Mohd Nurazzi**
Faculty of Engineering
Universiti Putra Malaysia
Serdang, Selangor, Malaysia

**M.T. Paridah**
Institute of Tropical Forestry and Forest Products
Universiti Putra Malaysia
Serdang, Selangor, Malaysia

**Md. Rezaur Rahman**
Department of Chemical Engineering
Universiti Malaysia Sarawak
Kota Samarahan, Sarawak, Malaysia

**J. Sahari**
Faculty of Science and Natural Resources
Universiti Malaysia Sabah
Kota Kinabalu, Sabah, Malaysia

**M.L. Sanyang**
University of The Gambia
The Gambia, West Africa

and

Universiti Putra Malaysia
Serdang, Selangor, Malaysia

# Contributors

**S.M. Sapuan**
Department of Mechanical and Manufacturing Engineering
Faculty of Engineering
and
Laboratory of Bio-Composite Technology
Institute of Tropical Forestry and Forest Products
Universiti Putra Malaysia
Serdang, Selangor, Malaysia

**R. Yahaya**
Science and Technology Research Institute for Defence
Kajang, Selangor, Malaysia

# 1 Natural Fiber Composites
## *Challenges and Opportunities*

*Faris M. Al-Oqla and S.M. Sapuan*

**CONTENTS**

1.1 Introduction ........................................................................................................1
1.2 Natural Fibers .....................................................................................................3
1.3 Life Cycle Assessments of NFCs .......................................................................4
1.4 Major Issues in the Development of NFCs ........................................................7
    1.4.1 Water Absorption Characteristics of NFCs ............................................7
    1.4.2 Compatibility of Fibers and Polymers in NFCs .....................................9
    1.4.3 Thermal Stability of Natural Fibers ......................................................11
    1.4.4 Factors Influence the Composite Performance ....................................14
1.5 Applications of NFCs .......................................................................................15
1.6 Future Developments .......................................................................................18
1.7 Summary ..........................................................................................................18
1.8 Conclusions ......................................................................................................18
References ................................................................................................................19

## 1.1 INTRODUCTION

Infrastructural physical components are usually constructed utilizing materials from finite resources such as steel, aluminum, and reinforced concrete. It is statistically proven that buildings alone consume about half of the total resources used globally. This has led to environmental damage and depletion of available natural resources. In addition, most of the conventional building-oriented materials and constructional processes are energy-intensive to produce, and are primarily responsible for a significant amount of landfill volume, thus producing about 40% of greenhouse gas emissions (Sallih, Lescher, and Bhattacharyya 2014). Furthermore, due to population growth, an increase in the demand for conventional materials leaves a large ecological footprint.

The growing interest in long-term sustainability, as well as awareness of environmental issues, has emphasized the proper utilization of natural resources by new environmental regulations. This has resulted in changing public and governmental attitudes and has stimulated considerable advancements in natural composite materials. Natural fiber composites (NFCs) have been recently highlighted in various industrial applications and have been slowly replacing conventional materials based upon several factors. (Abral et al. 2014; Agoudjil et al. 2011; Ahuja, Mir, and Kumar 2007; AL-Oqla and Sapuan 2014b; Almagableh, AL-Oqla, and Omari 2017).

Proper material selection has become pivotal in engineering to attain successful and sustainable design, as well as customer satisfaction attributes (AL-Oqla and Sapuan 2014b; Alves et al. 2010a). Moreover, the implementation of new materials in the industrial sector is usually limited by several constrains and limitations, such as the inherent relationship between the materials and their availability, cost, compatibility with the product design, machinability, recyclability, and performance in the final product form. This makes compromising these constraints, advantages, and disadvantages in selecting materials an intricate matter, where proper decisions have to be made concerning modern techniques like optimization methods, informative decisions, and expert systems utilizing the pairwise comparisons. (Dweiri and Al-Oqla 2006; AL-Oqla and Hayajneh 2007; Al-Oqla and Omar 2012; Al-Oqla and Omar 2014; Al-Widyan and Al-Oqla 2011, 2014; Dalalah, Al-Oqla, and Hayajneh 2010; Dieter 1997; Jahan 2010). In comparison to conventional composites, NFCs have greater specific strength and stiffness, better resistance to corrosion, greater fatigue strength and impact absorption capacities, recyclability, adaptability to hazardous environments, lower life-cycle costs, and non-toxicity (Dittenber and GangaRao 2011; Faruk et al. 2012a; AL-Oqla, Sapuan, Ishak, and Aziz 2014). Such advantages of NFCs resulted from the advantages of their constituents (fillers and polymers) particularly the natural fibers that have major advantages over traditional glass fibers. Such advantages include low cost, energy recovery, good thermal and acoustical insulation characteristics, availability, degradability, $CO_2$ sequestration enhancements, reduced dermal and respiratory irritation, and reduced tool wear in machining operations (Kalia 2011b; Faruk et al. 2012b; Alves et al. 2010b; Mir et al. 2010; Pickering et al. 2007; Sarikanat 2010). The features, as well as the performance of products, made from NFCs strongly depend upon the properties of their individual constituents and their compatibility as well as the polymer/filler interfacial characteristics that expand the possibilities of producing various exciting new materials with entirely new qualities (AL-Oqla, Sapuan et al. 2015; Al-Oqla and Sapuan 2015a).

The growing use of natural fiber reinforced polymer composites instead of synthetic fiber composites may provide even several long-term benefits to the overall sustainability, cleaner production theme, and infrastructure management (Al-Oqla, Sapuan, Ishak et al. 2015a, 2015b). However, there is uncertainty of performance associated with variability in natural fiber properties (AL-Oqla, Sapuan, Ishak et al. 2015b; AL-Oqla and Sapuan 2015b; AL-Oqla, Sapuan, Ishak, and Aziz 2014). This requires careful study of selecting the most high performance manufacturing for such types of composites under controlled conditions to achieve more reliable and better designed data (AL-Oqla, Sapuan, Ishak, and Aziz 2014; AL-Oqla, Sapuan, Ishak et al. 2015c).

## 1.2 NATURAL FIBERS

The use of agricultural raw material sources in the plastic industry would provide a renewable source of materials as well as generate non-food sources of economic development for several countries for long-term commercial development, where there must be a guaranteed long-term resource supply. Nature has offered humanity various types of natural fibers available in a wide range of colors, sizes, and shapes. Natural fibers can be classified regarding their origin as bast, leaf, fruit, and seed-hair fibers. This is illustrated in Figure 1.1. Various natural fiber types are available and suitable for natural fibers, including wood, bagasse, oil palm, pineapple leaf, date palm, cotton, rice straw, flax, hemp, rice husk, wheat straw, curaua, coir, doum fruit, ramie, jowar, kenaf, bamboo, sisal, rapeseed waste, and jute (Jawaid and Abdul Khalil 2011; AL-Oqla, Othman et al. 2014). Natural fibers are emerging as lightweight, available, low-cost, eco-friendly alternatives to glass fibers in composites.

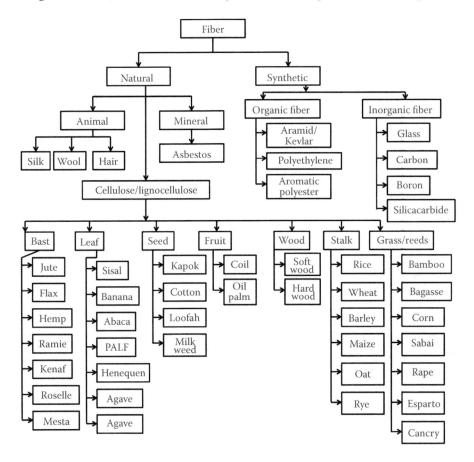

**FIGURE 1.1** Natural fibers and their classifications. (With kind permission from Springer Science+Business Media: *Biomass and Bioenergy*, Processing and properties of date palm fibers and its composites, 2014, 1–25, Al-Oqla, Faris M., Othman Y. Alothman, M. Jawaid, S.M. Sapuan, and M.H. Es-Saheb.)

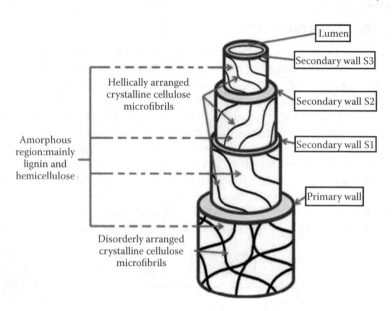

**FIGURE 1.2** Aschematic structure of a natural fiber. (Reprinted from *Materials & Design*, 47, Azwa, Z.N., B.F. Yousif, A.C. Manalo, and W. Karunasena, A review on the degradability of polymeric composites based on natural fibres, 424–442, Copyright 2013, with permission from Elsevier.)

Moreover, there are various types of natural fibers that are utilized as reinforcements in NFCs. Kenaf, date palm, hemp, coir, and flax are particularly attractive choices due to their rapid growth, availability over a wide range of climates, low cost, as well as their desired characteristics (AL-Oqla and Sapuan 2014a). For instance, a comparison of the specific moduli of natural and glass fibers relative to their costs per lengths in resisting 100 kN load demonstrates how kenaf, as well as other fibers, can compete with glass fibers (Joshi et al. 2004).

Natural fibers themselves can be considered as composites of hollow cellulose fibrils collected by lignin as a binder in a hemicellulose matrix. The cell wall is inhomogeneous with unique characteristics for each type of species of plant. A schematic structure of a natural fiber is demonstrated in Figure 1.2. The structure of fibers is complex with layers consisting of a thin primary wall as the first layer lifted during cell growth surrounding a secondary wall consisting of three distinct layers. The thickness of the middle layer usually regulates the fibers' ultimate mechanical properties. This middle layer consists of helically twisted cellular microfibrils (typically of 10–30 nm long) in a complex manner with long-chain cellulose molecules (organized of 30–100 cellulose molecules). Chemical compositions and microfibrillar angles of some natural fibers are presented in Table 1.1.

## 1.3 LIFE CYCLE ASSESSMENTS OF NFCs

Studies that compare life cycle assessments (LCAs) of natural and glass fiber composites have demonstrated and identified key drivers of natural fibers' attractive

### TABLE 1.1
### Chemical Compositions and Microfibrillar Angle of Some Natural Fibers

| Fiber Name | Cellulose [wt%] | Lignin [wt%] | Hemicellulose [wt%] | Pectin [wt%] | Wax [wt%] | Moisture [wt%] | Ash [wt%] | Microfibrillar Angle |
|---|---|---|---|---|---|---|---|---|
| Abaca | 56–63 | 7–9 | 20–25 | – | 3 | – | – | 20–25 |
| Bamboo | 26–43 | 1–31 | 30 | – | – | 9.16 | – | – |
| Coir | 37 | 42 | – | – | – | 11.36 | – | 30.45 |
| Cotton | 82.7–91 | – | 3 | – | 0.6 | 7.85–8.5 | – | – |
| Curaua | 73.6 | 7.5 | 9.9 | – | – | – | – | – |
| Flax | 64.1–71.9 | 2–2.2 | 64.1–71.9 | 1.8–2.3 | 1.7 | 8–1.2 | – | 5–10 |
| Hemp | 70.2–74.4 | 3.7–5.7 | 17.9–22.4 | 0.9 | 0.8 | 6.2–1.2 | 0.8 | 2–6.2 |
| Jute | 61–71.5 | 12–13 | 17.9–22.4 | 0.2 | 0.5 | 12.5–13.7 | 0.5–2 | 8 |
| Kenaf | 45–57 | 21.5 | 8–13 | 0.6 | 0.8 | 6.2–12 | 2–5 | 2–6.2 |
| Rachis | 42.75 | 26 | – | – | – | – | – | 28–37 |
| Ramie | 68.6–91 | 0.4–0.7 | 5–14.7 | 1.9 | – | – | – | 69–83 |
| Rice husk | 38–45 | – | 12–20 | – | – | – | 20 | – |
| Sea grass | 57 | 5 | 38 | 10 | – | – | – | – |
| Sisal | 78 | 8 | 10 | – | 2 | 11 | 1 | – |

environmental performances. LCAs are used for assessing the environmental features as well as potential impacts associated with a product, by gathering an inventory of related inputs and outputs of a product system and evaluating the possible environmental impacts related with those inputs and outputs. In addition, LCAs interpret the consequences of the inventory analysis and impact assessments related to the objectives of the study. The simplified, basic life cycle stages of a component made from natural fiber reinforced composite material is shown in Figure 1.3.

NFCs are likely to be environmentally superior to traditional glass fiber composites due to the following reasons:

1. The production of natural fibers has a lower environmental impact than the production of glass fibers.
2. NFCs have higher fiber content of that of glass fiber composites at an equivalent performance, reducing more polluting base polymer contents.
3. Lightweight NFCs reduce emissions in the use phase of the component and dramatically improve fuel efficiency, especially in automotive applications.
4. The natural fibers' end-of-life incineration leads to recovered energy and carbon credits.

A study considered life cycle assessments of a side panel for an Audi A3 car made from two different alternatives: a design made from ABS co-polymer and an alternative design made from a 66-volume hemp fiber epoxy resin composite that took energy use and emissions into consideration up to the component manufacturing

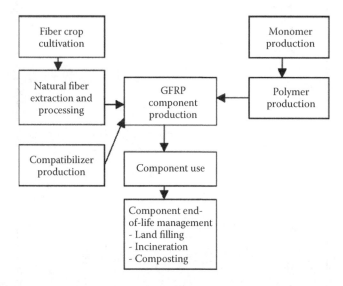

**FIGURE 1.3** Life cycle flowchart of a natural fiber reinforced composite component. (With kind permission from Springer Science+Business Media: *Composites Part A: Applied Science and Manufacturing*, Are natural fiber composites environmentally superior to glass fiber reinforced composites?, 35(3), 2004, 371–376, Joshi, Satish V., L.T. Drzal, A.K. Mohanty, and S. Arora.)

stage, as models inputs have demonstrated the superiority of NFCs over traditional synthetic fibers. Results of this study have proven that the natural fiber component uses 45% less energy, and results in lower air emissions. However, due to using fertilizer in hemp cultivation, water emissions of nitrates and phosphates besides nitrogen oxide (NOx) emissions to air were higher than that of ABS. Most of those studies showed that in their specific applications, NFCs are environmentally superior to synthetic fiber composites on most performance metrics. However, there were some differences across studies, in terms of the specific component/application being studied, the specific natural fiber chosen, the material composition of the reference component, the environmental impacts considered, the production processes, the boundaries and scope of the life cycle assessment, and the data sources used.

Therefore, it is necessary to establish some general drivers of the relative environmental performance of general fiber-reinforced polymer composites (natural and synthetic) that would allow us to make informed judgments without conducting exhaustive, time consuming, and expensive life cycle analyses every time. Such proposed methods may include elaborating on the more desirable characteristics of NFCs' constituents that designers have to take into consideration regarding the selection of NFC materials and their constituents to enhancing the performance of their designs. Also, researchers should conduct more comprehensive pairwise comparisons between different fiber types regarding some selective combined criteria to put emphasis on the need for better green and eco-friendly products (AL-Oqla, Sapuan, Ishak et al. 2015d; AL-Oqla, Sapuan, Ishak, and Nuraini 2014; AL-Oqla and Sapuan 2014b; AL-Oqla, Sapuan, Ishak et al. 2015c; AL-Oqla, Sapuan, Ishak, and Aziz 2014).

Regarding the matrix system, completely biodegradable materials are the most preferable materials to be used as matrices for the NFCs to fully satisfy regulations about using the eco-friendly and sustainable materials. However, the relative high cost of such types of materials is considered a disadvantage in comparison to commodity polymers, such as polypropylene, that can be easily recycled with a lower cost, exhibit stable performance behavior during their expected lifetime, and be capable of controlled degradation. The worldwide bio-based polymers production in various market segments in 2013 is illustrated in Figure 1.4.

The advantages as well as the disadvantages of the natural fiber reinforced polymer composites can be summarized in Table 1.2.

## 1.4 MAJOR ISSUES IN THE DEVELOPMENT OF NFCs

### 1.4.1 WATER ABSORPTION CHARACTERISTICS OF NFCs

During their lifetime, natural fiber-reinforced polymer composites are often offered to variable conditions, including the hygroscopic ones. The hydrophilic features of the natural fibers lead to a high level of moisture absorption in wet conditions, which causes a pronounced influence on their mechanical performance (AL-Oqla, Sapuan, Ishak, and Nuraini 2014). The hygroscopic characteristics of the bio-fibers is serious for their overall mechanical application, as they intensely influence the fiber-polymer bonding of NFCs, thus decreasing their desired mechanical performance (Sapuan et al. 2013; Azwa et al. 2013). Thus, the moisture absorption of natural fibers is one of

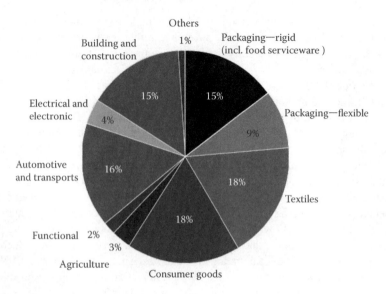

**FIGURE 1.4** Worldwide bio-based polymers production in various market segments in 2013. (From Al-Oqla, F.M. and Salit, M.S. 2017. *Materials Selection for Natural Fiber Composites*. Woodhead Publishing, Cambridge.)

**TABLE 1.2**
**Advantages and Disadvantages of NFCs**

| Advantages | Disadvantages |
|---|---|
| • Low densities | • Lower durability than synthetic fiber composites |
| • High specific strength and stiffness | |
| • Durability that can be improved considerably with treatment | • High moisture absorption, which usually results in performance deteriorations |
| • Fibers are from renewable resources | • Lower strength, particularly the impact strength compared to synthetic fiber composites |
| • Production requires little energy consumption | |
| • Production involves $CO_2$ absorption and returning oxygen to the environment | • Greater variability of properties |
| | • Lower processing temperatures limiting to matrix options |
| • Low-cost fibers | |
| • Low-hazard manufacturing processes | |
| • Low emission of toxic fumes at end-of-life | |
| • Less abrasive damage to processing equipment | |
| • More recyclability features | |

their major undesirable features, as it causes swelling, accelerating of fiber degradation, loss of mechanical properties, cracking, and breeding decaying fungi (Azwa et al. 2013; Sapuan et al. 2013).

In spite of that, the results in the literature regarding the effect of moisture on the fibers' performance were not consistent altogether (Célino et al. 2014).

It was reported that the amount of moisture absorption may enhance or worsen the composite properties depending on the fiber and matrix types (Azwa et al. 2013; AL-Oqla, Sapuan, Ishak, and Nuraini 2014). It was also reported that the amount of water absorption would reduce a particular mechanical property for a certain fiber type, but enhance it for others (Symington et al. 2009; Placet, Cisse, and Boubakar 2012; Azwa et al. 2013; AL-Oqla, Sapuan, Ishak, and Nuraini 2014). Inconsequently, the distinguished behavior of the fiber types regarding the effect of moisture is not well understood yet. This predicts that the relative performance of fibers under wet conditions during their life spans will be a difficult process. Besides, the results reported inconsistency regarding this effect of moisture makes the selection of an appropriate natural fiber type to form an NFC for unsteady conditions is a sophisticated process, which requires new methodologies to predict such effects on natural fibers (AL-Oqla, Sapuan, Ishak, and Nuraini 2014). Therefore, a novel systematic evaluation tool for predicting the natural fibers capabilities and performance based on the moisture content criterion (MCC) has been be developed and introduced as a new evaluation scheme (AL-Oqla, Sapuan, Ishak, and Nuraini 2014). This in order would dramatically contribute to the industrial sustainability theme by introducing new potential natural fibers types capable for supporting and growing the sustainable design possibilities. The flowchart of MCC methodology is shown in Figure 1.5.

The predicted performance of some natural fibers regarding the tensile modulus property after applying the MCC methodology shows that all of coir, date palm, and hemp fibers will have improved performance due to water absorption with a maximum of 8.6% for hemp, whereas both sisal and jute will be negatively affected, as seen in Figure 1.6. It was predicted that sisal will have about 13.5% reduction in its tensile modulus and 15.6% reduction for jute to be the worst performance compared to other fibers, which was inconsistent with previous published results in the literature as an indicator for the reliability of the MCC methodology.

### 1.4.2 Compatibility of Fibers and Polymers in NFCs

Natural fibers and their composites are not totally free of problems despite being competitive and having the advantages of low cost and low density over other fibers. As natural fibers have relatively strong polar bonds that usually cause a compatibility problem in bonding with most of the polymer matrices, various treatments processes surface of fibers can enhance the interfacial adhesion between the fiber and the polymer. However, this may increase the cost of natural fibers and decrease the water absorption of fibers. The interfacial bonding between fibers and polymers takes a vital role in governing the mechanical properties of NFCs (AL-Oqla 2017). As stress is conveyed between the polymer and fibers across the boundary, good interfacial bonding is essential to attain the most favorable reinforcement conditions. Nevertheless, if the interface bonding is too strong, crack propagation may arise, which has the possibility to reduce the toughness and strength of the composites. However, for bio-based composites there is usually inadequate interface between the hydrophilic fibers and matrices (as they are normally hydrophobic in nature) leading to a poor interfacial bonding that limits the NFC's mechanical performance (Al-Oqla, Sapuan, Ishak et al. 2015a; AL-Oqla, Sapuan, Ishak et al. 2015c; El-Sabbagh 2014). For bonding to occur, the natural fibers

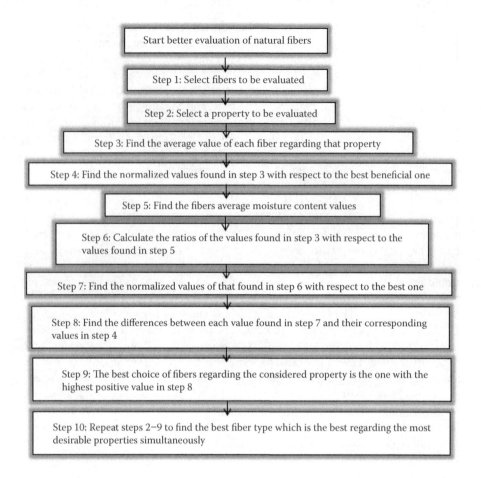

**FIGURE 1.5** The flowchart of MCC methodology. (From AL-Oqla, F.M., Sapuan, S., Ishak, M.R., and Nuraini, A.A. 2014. *BioResources*, 10 (1):299–312.)

and matrices must be brought into intimate contact, where wettability can be considered as an important sign in bonding. In addition, insufficient fiber wetting usually results in interfacial defects that can act as stress concentrators (Chen et al. 2006). Consequently, the fiber's wettability was reported to enhance the mechanical properties of bio-based composites like toughness, tensile, and flexural strengths of composites.

Chemical treatments therefore, may enhance and modify the characteristics of natural fibers so they are compatible for NFCs. Physical as well as chemical treatments can enhance the wettability of the fiber and thus improve the interfacial strength (Sinha and Panigrahi 2009; Asumani, Reid, and Paskaramoorthy 2012).

Some of the chemical treatments for natural fibers include the following (AL-Oqla, Alothman et al. 2014).

1. Alkaline treatment
2. Silane treatment

# Natural Fiber Composites

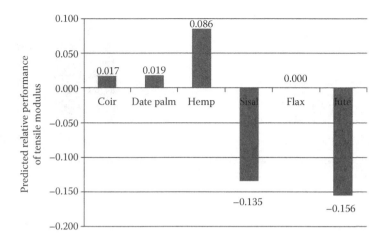

**FIGURE 1.6** Predicting the relative performance of tensile moduli for some fibers due to water absorption utilizing MCC methodology. (From AL-Oqla, F.M., Sapuan, S., Ishak, M.R., and Nuraini, A.A. 2014. *BioResources*, 10 (1):299–312.)

3. Maleated coupling agents
4. Benzoylation treatment
5. Peroxide treatment
6. Acetylation of natural fibers
7. Acrylation, maleic anhydride, and titanate treatment of natural fibers
8. Acrylation and acrylonitrile grafting
9. Isocyanate treatments
10. Etherification of natural fibers
11. Permanganate treatment
12. Sodium chlorite treatment of natural fibers
13. Plasma treatment

It is worthy to note here that the proper treatment conditions, like the proper solution type, concentration, and time would enhance the mechanical properties of the fibers. The effect of the chemical treatment of the date palm fiber on its mechanical properties for instance, is shown in Figure 1.7, where proper treatments obviously enhanced the tensile strength as well as the Young's modulus of the fiber. Moreover, the stress/strain diagrams of date palm fibers treated with different NaOH concentration are demonstrated in Figure 1.8.

## 1.4.3 Thermal Stability of Natural Fibers

Natural fibers are complex mixtures of organic materials, thus thermal treatment usually leads to several physical and chemical changes. The thermal stability of fibers is an important issue for NFCs and can be studied by thermogravimetric analysis (TGA). It was reported that the mechanical properties of natural fibers could be deteriorated as a result of thermal degradation particularly the toughness and bending

**12**  Kenaf Fibers and Composites

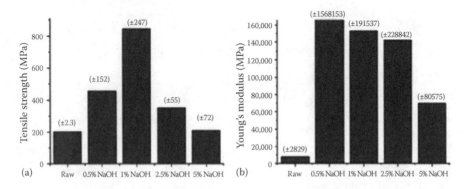

**FIGURE 1.7** The effect of NaOH treatments on the mechanical properties of date palm fibers (a) the tensile strength and (b) Young's modulus. (With kind permission from Springer Science+Business Media: *Green Biocomposites*, Design and fabrication of green biocomposites, 2017, 45–67, Al-Oqla, Faris M., Ahmad Almagableh, and Mohammad A. Omari.)

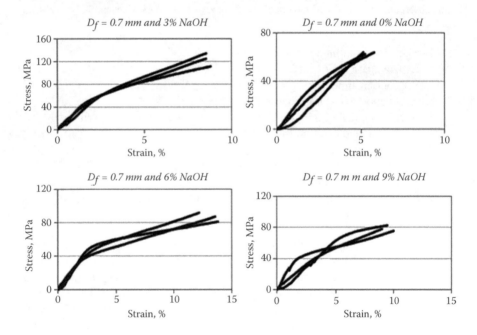

**FIGURE 1.8** Stress/strain diagram of date palm fiber after treated with different NaOH concentration. (From Shalwan, A., and B.F. Yousif, *Materials & Design* 53:928–937, 2014. doi: 10.1016/j.matdes.2013.07.083.)

strength (Sapuan, Haniffah, and AL-Oqla 2016; AL-Oqla 2017). It was also detected that thermal effects can change the surface chemistry and cause changes in the fiber/polymer bonding that is responsible for the inferior properties of NFCs. Moreover, the thermal degradation of the fibers also results in production of volatiles at processing temperatures, which usually lead to porous polymer products with

# Natural Fiber Composites

lower densities poor mechanical properties. This results in the fast biodegradability of green composites during fabrication processes.

For instance, the thermal characteristics of raw, ripe, and matured betel nut husk (BNH) fiber are demonstrated in Figures 1.9 and 1.10. They show the TGA results of the weight loss and the temperature at 5% weight loss for the fibers obtained at a

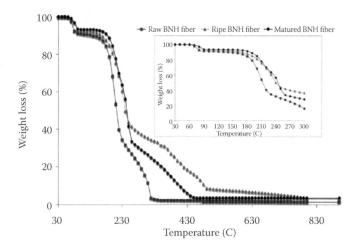

**FIGURE 1.9** TGA of raw, ripe, and matured BNH fiber. (Reprinted from *Journal of Cleaner Production*, 72, Yusriah, L., S.M. Sapuan, Edi Syams Zainudin, and M. Mariatti, Characterization of physical, mechanical, thermal and morphological properties of agro-waste betel nut (Areca catechu) husk fibre, 174–180, Copyright 2014, with permission from Elsevier.)

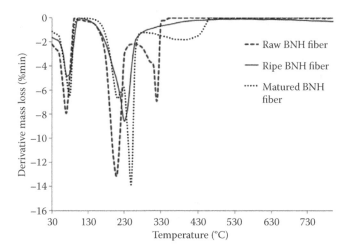

**FIGURE 1.10** DTG curves of BNH fibers. (Reprinted from *Journal of Cleaner Production*, 72, Yusriah, L., S.M. Sapuan, Edi Syams Zainudin, and M. Mariatti, Characterization of physical, mechanical, thermal and morphological properties of agro-waste betel nut (Areca catechu) husk fibre, 174–180, Copyright 2014, with permission from Elsevier.)

heating rate of 10°C/min and the initial degradation temperature was measured by the temperature at 5% degradation (Yusriah et al. 2014).

Based on the TGA curves of Figure 1.9 and the derivative of this thermogram (DTG) curves in Figure 1.10, the BNH fibers showed different thermal behavior. The three main degradation phases in the TGA curves are in Figure 1.9. The initial phase corresponds to the loss of moisture and highly volatile extractives from the fiber. It was located between 70 and 110°C. The second phase of degradation was due to the decomposition of hemicelluloses and occurred between 210 and 250°C. The last degradation phase corresponds to the decomposition of the lignin and occurred between 230 and 460°C. The DTG, on the other hand, highlights the inflection points of reactions that occurred in the entire degradation temperature range as illustrated in Figure 1.10. In this case, the first mass loss BNH fiber occurred between 35 and 100°C, with an endothermic peak at 74°C due to the dehydration process, followed by a second thermal decomposition occurring at an endothermic peak at 210°C, and another at 315°C corresponding to various mass losses in the fiber.

Generally, it was reported that the thermal stabilities of natural fibers are influenced by the type of fibers due to their different chemical compositions and the amounts of cellulose, hemicellulose, lignin, and moisture (Yusriah et al. 2014).

### 1.4.4 Factors Influence the Composite Performance

Functional requirements can restrict the design of components made of composites. As natural fibers would achieve the desired specific requirements of the composites (the proper selection of the reinforcement and the polymer on the one hand, and determining other technical aspects of the composites on the other), NFCs' constituents should be fairly executed and optimized. In fact, there are several criteria that may limit and affect the final desired characteristics of NFCs. Great efforts have been done by Al-Oqla and Sapuan (AL-Oqla and Sapuan 2014b) to classify these criteria into the following distinguished levels:

1. Natural fiber level
2. Matrix level
3. Composite level
4. General performance level
5. Specific performance level

For the first level, various properties regarding natural fibers have to be addressed, whereas the second level concerns the properties of the polymer matrix. The third level, on the other hand, considers properties concerning the composite itself, as the characteristics of the final composites are not parallel to any of these fibers or matrices. The general performance level involves different assets regarding the composite performance such as mechanical properties, weather resistance, biostability, life cycle, etc. The specific performance level, on the other hand, takes the particular requirements regarding a specific desired function or application into account. For automotive applications, all of the composites' weight, thermal

## TABLE 1.3
### Classified Criteria Regarding the Composite Level

| Level 1 Category | Level 2 Property/ Characteristic | Level 3 Criteria |
|---|---|---|
| Composite Characteristics | *Physical* | Surface topology, surface roughness, total density, texture, coefficient of thermal expansion, specific heat, electrical conductivity, reflective index, color and aesthetic, and opacity and translucency |
| | *Chemical and Biological* | Biostability, recyclability, biodegradability behavior, life cycle time, toxicity, weather resistance, storage (on-shelf storage), water absorption behavior, sunlight and UV resistance, and possibility of thermal recycling |
| | *Mechanical/Structural* | Yield strength, compressive strength, elastic modulus, shear modulus, impact strength, fatigue strength, flexural modulus, elongation to break, Poisson's ratio, fracture toughness, creep resistance, and hardness |
| | *Technical* | Sterilizability, fabrication knowledge and time, reproducibility, fabrication cost, product quality, packaging, thermal stability, level of automation, process parameters (pressure, temperature, cure time, and surface finish requirements), possibility of producing homogenous/nonhomogenous composites, secondary processability, labor protection and safety, life cycle cost, and cost of performance improvement |

properties, occupational health and safety, and acoustic insulation properties have to be considered.

The classified criteria that should be considered for the natural fiber composite level are shown in Table 1.3. According to these criteria, the composites of date palm fibers have the potential to be utilized in attractive applications, particularly in the automotive industry (AL-Oqla and Sapuan 2014b; Al-Oqla and Sapuan 2015a; AL-Oqla, Sapuan, Ishak et al. 2015b).

## 1.5 APPLICATIONS OF NFCs

Over the last couple of decades, a growing number of new car models encouraged by government regulations, in both Europe and North America, have featured natural, fiber-reinforced polymers in their components. Such components included door panels, instruments panels, hat racks, package trays, boot liners, internal engine covers, sun visors, and oil/air filters. Such composites are developed to be involved in more structurally demanding components like exterior underfloor paneling and seat backs. Nowadays, NFC materials are extensively used by all the main international automotive manufacturers (Faruk et al. 2014). The aircraft industry has also been embracing natural fiber-reinforced polymer composites for interior components (Ho et al. 2012). Furthermore, NFCs have been actually used in applications as a replacement for synthetic fiber composites as diverse as toys, cases for electronic devices (such as laptops and mobile phones), packaging, marine railings, and funeral articles (Ho et al. 2012). In sports, a number of companies are now presenting surfboards incorporating natural fiber-reinforced polymer composites. Recently, the production of natural fiber surfboards provides possibilities for better mechanical performance plus economic viability. Fishing rods are also being formed by means of materials developed by CelluComp Ltd. by extracting nanocellulose from root vegetables (Pickering 2008). Furthermore, some studies show that flax reinforced polyester turbine blades have the potential to replace blades made up from polymers reinforced with glass fibers (Shah 2013). Other researches have also shown the potential of implementing natural fiber composites for top plates of string musical instruments (Phillips and Lessard 2011). Besides, in the construction industry, wood fiber/polypropylene or fiber/polyethylene has been extensively utilized in decking, particularly in the United States. Natural fiber reinforced composites have also been gaining popularity in non-structural construction applications. They have been utilized for door and window frames, wall insulations, and floor lamination (Shah 2013). The replacement of wooden laminates in insulating structural panels has been assessed to find that NFCs have better mechanical properties. Reinforcement of cement by natural fibers for building materials was also assessed (Faruk et al. 2014). Overall, the worldwide natural fiber reinforced polymer composites market was estimated at US $2.1 billion in 2010 and estimated to rise 10% annually until 2016 (Ho et al. 2012) representative further potentials across wide range of industries including construction, automotive, aerospace, civil, and sports and leisure.

Several researchers have studied the appropriateness of the natural fibers as fillers for conductive polymers. Such works have examined the effects of several parameters like fiber length, fiber type, temperature, fiber content, humidity, and water absorption, and their chemical treatments to enhance their electrical characteristics. Such characteristics include dielectric constant, surface current curves, DC conductivity, electric charging phenomenon, capacitance, relaxation phenomenon, electrical resistivity, dielectric loss, volume and surface conductivities, dissipation and loss factors, and so on. It was testified that these natural fiber composites exhibit better electrical and thermal characteristics and may have potential applications as novel functional materials in textile and biological areas.

On the other hand, there is emerging attention in the expansion of the applications regarding electrically conductive polymeric composites, mainly, the natural fiber reinforced conductive polymer composites. Such interest arises as NFCs demonstrate mutual properties such as effective insulation and a high desirable mechanical strength that allows them to be admirable mechanical support for field carrying conductors. This enables these composites to be used in several areas such as printed circuit boards, terminals, insulators, connectors, switches, industrial plugs, and household plugs (Al-Oqla and Omar 2012; AL-Oqla and Omar 2015; AL-Oqla, Sapuan, Anwer et al. 2015). On the other hand, various textile fibers and fabrics such as cotton, viscose, rayon, polyester, and wool, are currently utilized with conducting polymers for modern applications like conductive fabrics, supercapacitors, electromagnetic interference, heating devices, shielding, and antimicrobial fabrics (Babu, Subramanian, and Kulandainathan 2013; Najar, Kaynak, and Foitzik 2007;

# Natural Fiber Composites

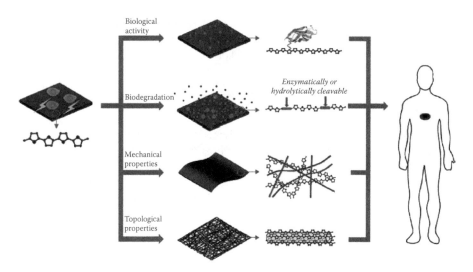

**FIGURE 1.11** Illustration of the aspects considered for biomimetic conducting polymer-based materials. (Reprinted from *Synthetic Metals*, 206, Al-Oqla, Faris M., S.M. Sapuan, T. Anwer, M. Jawaid, and M.E. Hoque, Natural fiber reinforced conductive polymer composites as functional materials: A review, 42–54, Copyright 2015, with permission from Elsevier.)

## TABLE 1.4
### Some Applications of Natural Fiber Reinforced Polymer Composites

| Industry | Application |
| --- | --- |
| Automotive | Seats, door panels, floor trays, insulation, rear parcel shelves, wheel shields, side and back door panels, seat backs, and hat racks |
| Building and Construction | Window frames, panels, door shutters, decking, railing systems, fencing, furniture industry cordages, construction products, textiles, cordages, geotextiles, manufacture of pipes, building panels, and bricks |
| Marine and Mechanical | Folded plates of various forms, both synclastic and anticlastic shells, skeletal structures, walls and panels, doors, windows, ladders, staircases, chemical and water tanks, cooling towers, bridge decks, antenna dishes, etc. |
| Aerospace | Wing skins, front fuselages, control surface fins and rudders, access doors, undercarriage doors, engine cowlings, main torsion boxes, fuel, floor boards, interior decorative panels, partitions, cabin baggage racks, and several similar applications |
| Sports | Skis, ski poles, fishing rods, golf clubs, tennis and badminton rackets, hockey sticks, poles (pole vault), bicycle frames, etc. |
| Others | Textiles and yarn, goods, paper and packaging, electrical, manufacture bank notes, furniture panels, constructing drains, and pipelines |

Hardy, Lee, and Schmidt 2013). A schematic diagram of the reflected aspects for designing biomimetic conducting polymer-based materials is shown in Figure 1.11. Some applications of natural fiber reinforced polymer composites in various industries are shown in Table 1.4.

## 1.6 FUTURE DEVELOPMENTS

Improvements in natural fiber reinforced polymer composites has occurred as a result of the enhancement of fiber selection as well as polymer-fiber evaluation methodologies, treatment, and interfacial engineering in addition to composite processing. Therefore, it is necessary to establish some general drivers of the relative environmental performance of general fiber reinforced polymer composites (natural and synthetic) that would allow us to make informed judgments without conducting exhaustive, time consuming and expensive life cycle analyses every time. Such proposed methods may include elaborating more desired characteristics of NFCs' constituents that designers have to take into consideration regarding selecting NFC materials and their constituents to achieve better performance in their designs.

## 1.7 SUMMARY

The growing interest in long-term sustainability, in addition to the awareness of environmental issues, the proper utilization of natural resources by new environmental regulations has been stressed. Consequently, NFCs have been highlighted in order to be implemented in various industrial applications and have been slowly replacing conventional materials. In comparison to conventional composites, NFCs have greater specific strength and stiffness, better resistance to corrosion, greater fatigue strength and impact absorption capacity, recyclability, adaptability to hazardous environments, lower life cycle costs, and non-toxicity. Such composites are developed to be involved to more structurally demanding components like exterior underfloor paneling and seat backs. Nowadays, NFCs materials are extensively used by all main international automotive manufacturers. In addition, they have been implemented in various applications as a replacement for synthetic fiber composites such as toys, cases for electronic devices, laptops and mobile phones, packaging, marine railings, and other applications.

## 1.8 CONCLUSIONS

The interfacial bonding between fibers and polymers takes a vital role in governing the mechanical properties of NFCs. As stress is conveyed between polymer and fibers across the boundary, good interfacial bonding is essential to attain the most favorable reinforcement conditions. However, for bio-based composites there is usually inadequate interface between the hydrophilic fibers and matrices (as they are normally hydrophobic in nature) leading to a poor interfacial bonding that limits the NFC's mechanical performance. Consequently, NFCs and their constituents must be properly evaluated regarding wide range of criteria so that to optimize their desired performance and extend their usage into wider modern applications.

## REFERENCES

Abral, H., D. Kadriadi, A. Rodianus, P. Mastariyanto, S. Arief, Mohd Salit Sapuan, and Mohamad Ridzwan Ishak. 2014. "Mechanical properties of water hyacinth fibers—Polyester composites before and after immersion in water." *Materials & Design* 58:125–129.

Agoudjil, Boudjemaa, Adel Benchabane, Abderrahim Boudenne, Laurent Ibos, and Magali Fois. 2011. "Renewable materials to reduce building heat loss: Characterization of date palm wood." *Energy and Buildings* 43 (2):491–497.

Ahuja, Tarushee, Irfan Ahmad Mir, and Devendra Kumar. 2007. "Biomolecular immobilization on conducting polymers for biosensing applications." *Biomaterials* 28 (5):791–805.

AL-Oqla, Faris M., Mohd Salit Sapuan, Mohamad Ridzwan Ishak, and Nuraini Abdul Aziz. 2015a. "Decision-making model for optimal reinforcement condition of natural fiber composites." *Fibers and Polymers* 16 (1):153–163.

AL-Oqla, Faris M., Mohd Salit Sapuan, Mohamad Ridzwan Ishak, and Nuraini Abdul Aziz. 2015b. "Selecting natural fibers for bio-based materials with conflicting criteria." *American Journal of Applied Sciences* 12 (1):64–71.

AL-Oqla, Faris M. and M.T. Hayajneh. 2007. "A design decision-making support model for selecting suitable product color to increase probability." Design Challenge Conference: Managing Creativity, Innovation, and Entrepreneurship, Amman, Jordan.

Al-Oqla, Faris M. and A.A. Omar. 2012. "A decision-making model for selecting the GSM mobile phone antenna in the design phase to increase over all performance." *Progress in Electromagnetics Research C* 25:249–269. doi: 10.2528/PIERC11102702.

AL-Oqla, Faris M. 2017. "Investigating the mechanical performance deterioration of Mediterranean cellulosic cypress and pine/polyethylene composites." *Cellulose*. doi: 10.1007/s10570-017-1280-3.

AL-Oqla, Faris M., Ahmad Almagableh, and Mohammad A. Omari. 2017. "Design and Fabrication of Green Biocomposites." *Green Biocomposites*, 45–67. Cham, Switzerland: Springer.

AL-Oqla, Faris M., Othman Y. Alothman, M. Jawaid, S.M. Sapuan, and M.H. Es-Saheb. 2014. "Processing and Properties of Date Palm Fibers and Its Composites." *Biomass and Bioenergy*, 1–25. Cham, Switzerland: Springer.

Al-Oqla, Faris M. and Amjad A. Omar. 2014. "An Expert-based model for selecting the most suitable substrate material type for antenna circuits." *International Journal of Electronics* (Just accepted).

AL-Oqla, Faris M. and Amjad A. Omar. 2015. "An expert-based model for selecting the most suitable substrate material type for antenna circuits." *International Journal of Electronics* 102 (6):1044–1055.

AL-Oqla, Faris M. and Mohd Sapuan Salit. 2017. "Materials Selection for Natural Fiber Composites." Cambridge: Woodhead Publishing.

AL-Oqla, Faris M., Mohd Salit Sapuan, Mohamad Ridzwan Ishak, and Nuraini Abdul Aziz. 2015. "Selecting Natural Fibers for Industrial Applications." Postgraduate Symposium on Biocomposite Technology Serdang, Malaysia, March, 3, 2015.

Al-Oqla, Faris M., and Mohd Salit Sapuan. 2015a. "Polymer selection approach for commonly and uncommonly used natural fibers under uncertainty environments." *Journal of the Minerals Metals and Materials Society*. doi: 10.1007/s11837-015-1548-8.

AL-Oqla, Faris M., Mohd Salit Sapuan, Mohamad Ridzwan Ishak, and Nuraini Abdul Aziz. 2014. "Combined multi-criteria evaluation stage technique as an agro waste evaluation indicator for polymeric composites: Date palm fibers as a case study." *BioResources* 9 (3):4608–4621. doi: 10.15376/biores.9.3.4608-4621.

AL-Oqla, Faris M., and Mohd Salit Sapuan. 2015b. "Polymer Selection Approach for Commonly and Uncommonly Used Natural Fibers Under Uncertainty Environments." *JOM* 67 (10):2450–2463.

AL-Oqla, Faris M., Mohd Salit Sapuan, T. Anwer, M. Jawaid, and M.E. Hoque. 2015. "Natural fiber reinforced conductive polymer composites as functional materials: A review." *Synthetic Metals* 206:42–54.

AL-Oqla, Faris M., S.M. Sapuan, Mohamad Ridzwan Ishak, and Abdul Aziz Nuraini. 2014. "A novel evaluation tool for enhancing the selection of natural fibers for polymeric composites based on fiber moisture content criterion." *BioResources* 10 (1):299–312.

AL-Oqla, Faris M., Mohd Salit Sapuan, Mohamad Ridzwan Ishak, and Nuraini Abdul Aziz. 2015c. "A Model for Evaluating and Determining the Most Appropriate Polymer Matrix Type for Natural Fiber Composites." *International Journal of Polymer Analysis and Characterization* 20 (Just accepted):191–205.

AL-Oqla, Faris M., Mohd Salit Sapuan, M.R. Ishak, and A.A. Nuraini. 2015d. "Predicting the potential of agro waste fibers for sustainable automotive industry using a decision making model." *Computers and Electronics in Agriculture* 113:116–127.

AL-Oqla, Faris M., and Mohd Salit Sapuan. 2014a. "Date Palm Fibers and Natural Composites." Postgraduate Symposium on Composites Science and Technology 2014 & 4th Postgraduate Seminar on Natural Fibre Composites 2014, 28/01/2014, Putrajaya, Selangor, Malaysia.

AL-Oqla, Faris M., and Mohd Salit Sapuan. 2014b. "Natural fiber reinforced polymer composites in industrial applications: Feasibility of date palm fibers for sustainable automotive industry." *Journal of Cleaner Production* 66:347–354. doi: 10.1016/j.jclepro.2013.10.050.

Al-Widyan, Momamad I., and Faris M. Al-Oqla. 2011. "Utilization of supplementary energy sources for cooling in hot arid regions via decision-making model." *International Journal of Engineering Research and Applications* 1 (4):1610–1622.

Al-Widyan, Momamad I., and Faris M. Al-Oqla. 2014. "Selecting the most appropriate corrective actions for energy saving in existing buildings A/C in hot arid regions." *Building Simulation* 7 (5):537–545. doi: DOI 10.1007/s12273-013-0170-3.

Almagableh, Ahmad, Faris M. AL-Oqla, and Mohammad A. Omari. 2017. "Predicting the Effect of Nano-Structural Parameters on the Elastic Properties of Carbon Nanotube-Polymeric based Composites." *International Journal of Performability Engineering* 13 (1):73.

Alves, C., P.M.C. Ferrão, A.J. Silva, L.G. Reis, M. Freitas, L.B. Rodrigues, and D.E. Alves. 2010a. "Ecodesign of automotive components making use of natural jute fiber composites." *Journal of Cleaner Production* 18 (4):313–327. doi: 10.1016/j.jclepro.2009.10.022.

Alves, C., P.M.C. Ferrão, A.J. Silva, L.G. Reis, M. Freitas, L.B. Rodrigues, and D.E. Alves. 2010b. "Ecodesign of automotive components making use of natural jute fiber composites." *Journal of Cleaner Production* 18 (4):313–327. doi: http://dx.doi.org/10.1016/j.jclepro.2009.10.022.

Asumani, O.M.L., R.G. Reid, and R. Paskaramoorthy. 2012. "The effects of alkali–silane treatment on the tensile and flexural properties of short fibre nonwoven kenaf reinforced polypropylene composites." *Composites Part A: Applied Science and Manufacturing* 43 (9):1431–1440.

Azwa, Z.N., B.F. Yousif, A.C. Manalo, and W. Karunasena. 2013. "A review on the degradability of polymeric composites based on natural fibres." *Materials & Design* 47:424–442.

Babu, K. Firoz, S.P. Siva Subramanian, and M. Anbu Kulandainathan. 2013. "Functionalisation of fabrics with conducting polymer for tuning capacitance and fabrication of supercapacitor." *Carbohydrate Polymers* 94 (1):487–495.

Célino, Amandine, Sylvain Fréour, Frédéric Jacquemin, and Pascal Casari. 2014. "The hygroscopic behavior of plant fibers: A review." *Frontiers in Chemistry* 1:1–12. doi: 10.3389/fchem.2013.00043.

Chen, Ping, Chun Lu, Qi Yu, Yu Gao, Jianfeng Li, and Xinglin Li. 2006. "Influence of fiber wettability on the interfacial adhesion of continuous fiber-reinforced PPESK composite." *Journal of Applied Polymer Science* 102 (3):2544–2551.

Dalalah, D., F.M. Al-Oqla, and M. Hayajneh. 2010. "Application of the Analytic Hierarchy Process (AHP) in multi-criteria analysis of the selection of cranes." *Jordan Journal of Mechanical and Industrial Engineering, JJMIE* 4 (5):567–578.

Dieter, George E. 1997. *ASM Handbook, Materials Selection and Design.* Vol. 20: ASM International.

Dittenber, David B., and Hota V.S. GangaRao. 2011. "Critical review of recent publications on use of natural composites in infrastructure." *Composites Part A: Applied Science and Manufacturing* 43 (8):1419–1429.

Dweiri, F., and F.M. Al-Oqla. 2006. "Material selection using analytical hierarchy process." *International journal of computer applications in technology* 26 (4):182–189. doi: 10.1504/IJCAT.2006.010763.

El-Sabbagh, A. 2014. "Effect of coupling agent on natural fibre in natural fibre/polypropylene composites on mechanical and thermal behaviour." *Composites Part B: Engineering* 57:126–135.

Faruk, Omar, Andrzej K. Bledzki, Hans-Peter Fink, and Mohini Sain. 2012a. "Biocomposites reinforced with natural fibers: 2000–2010." *Progress in Polymer Science* 37:1552–1596.

Faruk, Omar, Andrzej K. Bledzki, Hans-Peter Fink, and Mohini Sain. 2012b. "Biocomposites reinforced with natural fibers: 2000–2010." *Progress in Polymer Science* 37 (11):1552–1596. doi: http://dx.doi.org/10.1016/j.progpolymsci.2012.04.003.

Faruk, Omar, Andrzej K. Bledzki, Hans-Peter Fink, and Mohini Sain. 2014. "Progress Report on Natural Fiber Reinforced Composites." *Macromolecular Materials and Engineering* 299 (1):9–26.

Hardy, John G., Jae Y. Lee, and Christine E. Schmidt. 2013. "Biomimetic conducting polymer-based tissue scaffolds." *Current Opinion in Biotechnology* 24 (5):847–854.

Ho, Mei-Po, Hao Wang, Joong-Hee Lee, Chun-Kit Ho, Kin-Tak Lau, Jinsong Leng, and David Hui. 2012. "Critical factors on manufacturing processes of natural fibre composites." *Composites Part B: Engineering* 43 (8):3549–3562.

Jahan A., M.Y. Ismail M.Y., S.M. Sapuan, and F. Mustapha. 2010. "Material screening and choosing methods—A review." *Materials and Design* 31:696–705.

Jawaid, M.H.P.S. and H.P.S. Abdul Khalil. 2011. "Cellulosic/synthetic fibre reinforced polymer hybrid composites: A review." *Carbohydrate Polymers* 86 (1):1–18.

Joshi, Satish V., L.T. Drzal, A.K. Mohanty, and S. Arora. 2004. "Are natural fiber composites environmentally superior to glass fiber reinforced composites?" *Composites Part A: Applied Science and Manufacturing* 35 (3):371–376.

Kalia, Susheel, Alain Dufresne, Bibin Mathew Cherian, BS Kaith, Luc Avérous, James Njuguna, and Elias Nassiopoulos. 2011. Cellulose-based bio-and nanocomposites: A review. *International Journal of Polymer Science* 2011:1–35.

Mir, Abdallah, Redouane Zitoune, Francis Collombet, and Boudjema Bezzazi. 2010. "Study of mechanical and thermomechanical properties of jute/epoxy composite laminate." *Journal of Reinforced Plastics and Composites* 29 (11):1669–1680.

Najar, Saeed Shaikhzadeh, Akif Kaynak, and Richard C. Foitzik. 2007. "Conductive wool yarns by continuous vapour phase polymerization of pyrrole." *Synthetic Metals* 157 (1):1–4.

Phillips, Steven, and Larry Lessard. 2011. "Application of natural fiber composites to musical instrument top plates." *Journal of Composite Materials*: 0021998311410497.

Pickering, Kim. 2008. *Properties and Performance of Natural-Fibre Composites.* United Kingdom: Wookhead; North America: CRC Press.

Pickering, K.L., G.W. Beckermann, S.N. Alam, and N.J. Foreman. 2007. "Optimising industrial hemp fibre for composites." *Composites Part A: Applied Science and Manufacturing* 38 (2):461–468.

Placet, Vincent, Ousseynou Cisse, and M. Lamine Boubakar. 2012. "Influence of environmental relative humidity on the tensile and rotational behaviour of hemp fibres." *Journal of Materials Science* 47 (7):3435–3446.

Sallih, Nabihah, Peter Lescher, and Debes Bhattacharyya. 2014. "Factorial study of material and process parameters on the mechanical properties of extruded kenaf fibre/polypropylene composite sheets." *Composites Part A: Applied Science and Manufacturing* 61:91–107.

Sapuan, S.M., W.H. Haniffah, and Faris M. Al-Oqla. 2016. "Effects of Reinforcing Elements on the Performance of Laser Transmission Welding Process in Polymer Composites: A Systematic Review." *International Journal of Performability Engineering* 12 (6):553.

Sapuan, S.M., Fei-ling Pua, Y.A. El-Shekeil, and Faris M. Al-Oqla. 2013. "Mechanical properties of soil buried kenaf fibre reinforced thermoplastic polyurethane composites." *Materials & Design* 50: 467–470. doi: 10.1016/j.matdes.2013.03.013.

Sarikanat, Mehmet. 2010. "The influence of oligomeric siloxane concentration on the mechanical behaviors of alkalized jute/modified epoxy composites." *Journal of Reinforced Plastics and Composites* 29 (6):807–817.

Shah, Darshil U. 2013. "Developing plant fibre composites for structural applications by optimising composite parameters: A critical review." *Journal of Materials Science* 48 (18):6083–6107.

Shalwan, A., and B.F. Yousif. 2014. "Investigation on interfacial adhesion of date palm/epoxy using fragmentation technique." *Materials & Design* 53:928–937. doi: 10.1016/j.matdes.2013.07.083.

Sinha, Ela, and S. Panigrahi. 2009. "Effect of plasma treatment on structure, wettability of jute fiber and flexural strength of its composite." *Journal of composite materials* 43 (17):1791–1802.

Symington, Mark C., William M. Banks, David West, and R.A. Pethrick. 2009. "Tensile testing of cellulose-based natural fibers for structural composite applications." *Journal of Composite Materials* 43 (9):1083–1108.

Yusriah, L., S.M. Sapuan, Edi Syams Zainudin, and M. Mariatti. 2014. "Characterization of physical, mechanical, thermal and morphological properties of agro-waste betel nut (*Areca catechu*) husk fibre." *Journal of Cleaner Production* 72:174–180.

# 2 Kenaf Fiber
## *Structure and Properties*

A.H. Juliana, H.A. Aisyah, M.T. Paridah,
C.C.Y. Adrian, and S.H. Lee

### CONTENTS

2.1 Introduction ..........................................................................................................23
2.2 Kenaf .....................................................................................................................24
2.3 Kenaf Stems ..........................................................................................................24
2.4 Anatomical Properties of Kenaf Fibers ................................................................24
2.5 Chemical Properties of Kenaf Fibers ...................................................................28
2.6 Physical Properties of Kenaf Fibers .....................................................................30
2.7 Mechanical Properties of Kenaf Fibers ................................................................30
2.8 Summary ...............................................................................................................32
References ....................................................................................................................33

### 2.1 INTRODUCTION

Natural fibers are elongated substances produced by plants and animals. They offer a low-cost alternative and are lower in density compared to synthetic fibers. Nowadays, natural fibers, especially plant fibers, have attracted the attention of many researchers. Plant fibers can be obtained from various parts of plants, namely the stems (kenaf (mesta), jute, flax, hemp, ramie, and bamboo), leaves (abaca, sisal, and manila), seeds (cotton and kapok), fruit (coir), and other grass fibers. Plant fibers consist of cellulose, hemicelluloses, and lignin, where the individual percentage of these components varies with the different types of fibers and is also affected by the origin, growing, and harvesting conditions (Blackburn 2005; Westman et al. 2010).

Plant fibers are sclerenchyma elongated cells and are arranged longitudinally. The cells are long and narrowed at the cell ends and are surrounded and protected by a cell wall that is a complex macromolecular structure. Sclerenchyma gives mechanical strength and rigidity to the plant, since it is usually a supporting tissue in plants. Fibers are also associated with the xylem and phloem tissues of monocotyledonous and dicotyledonous plant stems and leaves (Smole et al. 2013). As reported by Farnfield (1975) and Vincent (2000), bast, fruit, and leaf fibers are naturally organized into bundles, and are therefore called fiber bundles, whereas fibers originating from seeds are single cells and are referred to as fibers. The exact number of ultimate fibers in one bundle is not known because it is not possible to model the separation processes to produce a specific bundle diameter. Decortication and retting techniques are commonly used to separate the fiber bundles from the leaves and bast of fiber plants (Mwaikambo 2006). Kenaf (*Hibiscus cannabinus* L.) is one of the prominent annual bast fiber, and is a promising fiber source for various applications such as composites, pulp and paper, insulation mats, absorbents, bedding material, and solid biofuel (Pande and Roy 1998; Azizi Mossello et al. 2010; Monti and Alexopolou 2013).

## 2.2 KENAF

Kenaf was introduced in Malaysia in the early 1970s and was recognized as a potential alternative fibrous material for the pulp, paper, and wood composite industries in the late 1990s under the 7th Malaysian Plan 1996–2000 (Abdul Khalil et al. 2010). Compared to other plants, the yields of kenaf are relatively higher (up to 25 t/ha) thus presenting more economic advantages (Wood 2000). For the past several years, kenaf fibers have been shown to be suitable for composite applications such as particleboard, MDF, wood plastic composite (WPC), non-woven materials, and pultruded products (Kawai 2005; Viilar et al. 2009; Dutt et al. 2009; Paridah et al. 2009a; Juliana et al. 2012; Aisyah et al. 2013).

The characteristics of kenaf fibers are similar to those of wood compared to hemp, flax, and jute fibers. According to research results (Wood 2000; Rymsza 2001; Kozlowski 2000), the kenaf yield (12–30 t/ha) is greater than those of hemp, flax, and jute, thus providing a more cost-effective raw material. Paridah and Khalina (2009b) reported that under a Malaysian climate, yields of kenaf vary from 2 t/ha to 25 t/ha.

## 2.3 KENAF STEMS

Kenaf is a woody-stemmed, herbaceous dicotyledon and can reach a height and stem diameter of 2.5–4.2 m and 25–51 mm, respectively (Rowell and Stout 1998; Azizi Mosello et al. 2010; Juliana 2013; Juliana et al. 2014). As shown in Figure 2.1, the kenaf tree is straight with a few branches. It is fast-growing plant and can be harvested in just 4–5 months after planting (Paridah and Khalina 2009b), thus it can be planted twice a year. The stem of kenaf comprises of two major parts; bast (30% w/w) and core (70% w/w), that are distinctly different in term of anatomy, physical, chemical, and mechanical properties; the fibrous outer bark or bast (Figure 2.2a) and woody inner core (Figure 2.2b). A clear illustration of the bast and core parts is shown in Figure 2.3. The fibers from these two parts differ greatly in terms of fiber morphology and chemical composition (Abdul Khalil et al. 2010; Azizi Mossello et al. 2010).

## 2.4 ANATOMICAL PROPERTIES OF KENAF FIBERS

Kenaf, *Tossa* jute, and white jute fibers are arranged in the bast or phloem region of the plant consisting of pyramidal wedges which taper outwards; the fiber bundles in each wedge are arranged in 8–24 layers, alternating with groups of thin-walled phloem (Sen 2009). Generally, the bast appear as long fibers, while the core contains short, woody

# Kenaf Fiber

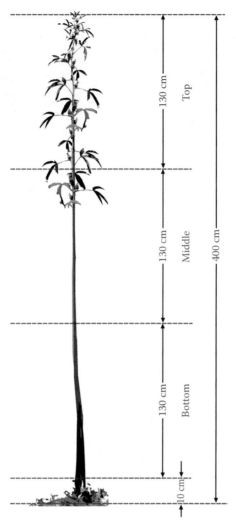

**FIGURE 2.1** Kenaf tree. (From Juliana, A.H. *Mechanical and Physical Properties of Particleboard Manufactured from Kenaf* (Hibiscus cannabinus L.) *and Rubberwood* (Hevea brasiliensis), School of Graduate Studies, Universiti Putra Malaysia, 1–237, 2013.)

fibers. The fibers of both the bast and core parts differ significantly in terms of their appearance, anatomical structure, and fiber morphology. Table 2.1 shows the comparison between fiber lengths and diameters of bast, core, softwood, and hardwood fibers. As tabulated in Table 2.1, kenaf bast fiber is slightly shorter and thinner compared to softwood fibers. This factor may increase the ability of bonding and strength development. On the other hand, fibers from the core are shorter (0.72 mm) and resemble hardwood fibers. Ashori (2006) and Abdul Khalil et al. (2010) reported that the length of kenaf bast fibers was in range of 2.5–3.6 mm, which also resemble those of softwood, meanwhile the core are shorter with an average value of 1.1 mm long. Besides that, according to Ververis et al. (2004), in a study of the fiber dimensions of

**FIGURE 2.2** (a) The outer bark and (b) the inner woody core of kenaf. (From Aisyah, H.A., *Kenaf* (Hibiscus Cannabinus L.) *Fibres Properties under Varied Thermo-Mechanical Pulping (TMP) Conditions and Their Influence on MDF Performance*, School of Graduate Studies, Universiti Putra Malaysia, 1–132, 2013.)

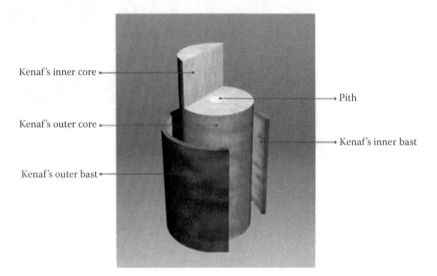

**FIGURE 2.3** Kenaf parts. (From Juliana, A.H. *Mechanical and Physical Properties of Particleboard Manufactured from Kenaf* (Hibiscus cannabinus L.) *and Rubberwood* (Hevea brasiliensis), School of Graduate Studies, Universiti Putra Malaysia, 1–237, 2013.)

annual plants and agricultural residues, they examined that the slenderness ratio (fiber length/fiber diameter) of kenaf bast fibers is comparable to those of softwood fibers such as cotton (*Gossypium hirsutum* L.) stalks, miscanthus (*Miscanthus giganteus*), and switchgrass (*Panicumvirgatum* L.).

A study done by Abdul Khalil et al. (2010) stated that the transverse sections of kenaf cell wall fibers are composed of intercellular layers: the primary wall and the secondary wall (S), which are made up of three layers (S1, S2, and S3). Core fibers showed great variability in size, shape, and structure of the cell wall fibers. In lignocellulosic fibers, these layers contain cellulose, hemicelluloses, and lignin, where the individual fibers are bonded together by a lignin-rich region known as

## TABLE 2.1
### Fiber Length and Diameter of Softwood, Hardwood, and Kenaf

| Fiber Comparison | Fiber Length (mm) Min | Max | Fiber Diameter (μm) Min | Max |
|---|---|---|---|---|
| Kenaf Bast[a] | 1.4 | 5.0 | 14 | 23 |
| Kenaf core[b] | 1.37 | 2.81 | 19 | 23 |
| Kenaf core[a] | 0.4 | 1.1 | 18 | 37 |
| Kenaf core[b] | 1.06 | 2.64 | 16 | 23 |
| Softwood | 2.7 | 4.6 | 32 | 43 |
| Hardwood | 0.7 | 1.6 | 20 | 30 |

*Source:* [a]Touzinsky, G.F. Kenaf, Secondary Fibers and Non-Wood Pulping, Chapter 8, in *Pulp and Paper Manufacturing*, Vol. 3. TAPPI Press, Atlanta, GA, 106, 1987; [b]Aisyah, H.A., *Kenaf (Hibiscus Cannabinus L.) Fibres Properties under Varied Thermo-Mechanical Pulping (TMP) Conditions and Their Influence on MDF Performance*, School of Graduate Studies, Universiti Putra Malaysia, 1–132, 2013.

middle lamella. Cellulose attains its highest concentration in the S2 layer, and lignin is most concentrated in the middle lamella. The S2 layer is usually the thickest layer and dominates the properties of the fibers. The S1 layer in core fibers is well-defined compared to bast fibers, and could be distinguished from the adjoining S2 layer (Figure 2.4). The S2 layers of core fibers occupies the largest area of all layers in the cell wall and has the thickest wall layer.

In terms of lumen diameter, apart from the thick cell wall, kenaf bast fibers have a wide variation of small lumen sizes compared to core fibers. This is well illustrated in Figure 2.5, which shows the cross section of kenaf bast (kenaf's outer bast and

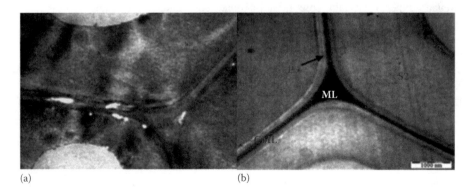

(a)  (b)

**FIGURE 2.4** Transverse section of (a) kenaf cell wall bast, and (b) kenaf cell wall core fibers. ML, middle lamella; P, primary wall; $S_1$, $S_2$ and $S_3$, secondary wall sub layers; CML, compound middle lamella. (From Abdul Khalil, H.P.S. et al., *Industrial Crops and Products* 31(1): 113–121, 2010.)

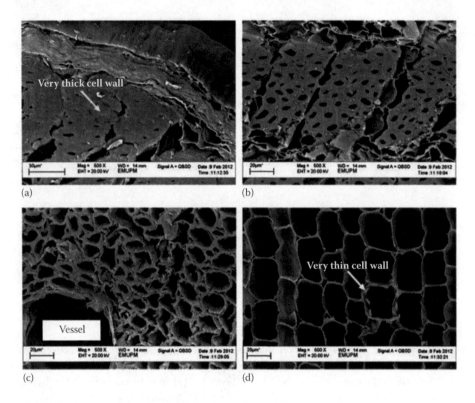

**FIGURE 2.5** A cross section between KOB, KIB, KOC, and KIC under 500x magnification using SEM. (a) KOB shows thickest cell wall, (b) KIB shows thick cell wall, (c) KOC shows thin cell wall (d) KIC shows very thin cell wall. KOB, Kenaf's' outer bast; KIB, Kenaf's' inner bast; KOC, Kenaf's' outer core; KIC, Kenaf's' inner core. (From Juliana, A.H. et al., Journal of Adhesion Science and Technology, 28(6), 546–560, 2014.)

kenaf's inner bast) and core (kenaf's outer core and kenaf's inner core). The microscopic appearance of kenaf core fibers are similar to those observed in many diffuse-porous hardwoods (Voulgaridis et al. 2000). The average lumen diameter of kenaf bast and core fibers was 2.8 μm and 6.7 μm, respectively.

Based on factors discussed above, bast fibers are able to provide high strength and can be converted to high performance composites (Ohnishi et al. 2003; Kawai et al. 1999). Meanwhile, core fibers have a potential for replacing wood and other lignocellulosic fibers especially in the pulp and paper industry (Abdul Khalil et al. 2010).

## 2.5 CHEMICAL PROPERTIES OF KENAF FIBERS

In general, lignocellulosic materials consists of cellulose, hemicelluloses, lignin, extractives, and inorganic matter. The bast and core of Kenaf are quite different from each other with respect to their chemical compositions. The chemical composition for both bast and core fibers is given in Table 2.2. The chemical composition of kenaf is quite similar to that of wood, except for the ash content, which is relatively higher.

## TABLE 2.2
## Chemical Composition of Kenaf Bast and Core Fibers

| Chemical Composition | Kenaf Bast[a] | Kenaf Core[a] | Softwood[b] | Hardwood[b] |
|---|---|---|---|---|
| Extractives (%) | 5.5 | 4.7 | 0.2–8.5 | 0.1–7.7 |
| Holocellulose (%) | 86.8 | 87.2 | 60–80 | 71–89 |
| α-Cellulose (%) | 55.0 | 49.0 | 30–60 | 31–64 |
| Lignin (%) | 14.7 | 19.2 | 21–37 | 14–34 |
| Ash (%) | 5.4 | 1.9 | <1 | <1 |

*Sources:* [a]Abdul Khalil, H.P.S. et al., *Industrial Crops and Products*, 31(1): 113–121, 2010; [b]Tsoumis, G., Science and Technology of Wood: Structure, *Properties and Utilization*. Van Nostrand Reinhold, New York, 1991.

In comparison, core fibers have higher percentages of holocellulose and lignin, while bast fibers have higher percentages of α-cellulose, extractives, and ash content. The α-cellulose content in the bast (55%) is higher than in the core (49%). The high α-cellulose content is believed to provide high strength in paper formation and other fiber-based end products (Abdul Khalil et al. 2010). Thi Bach et al. (2003) stated that during maturation, the cellulose (bast fiber 52–59%, core fiber 44–46%) and lignin (bast fiber 9.3–13.2%, core fiber 18.3–23.2%) contents increased significantly. The lignin content in kenaf fibers is lower than softwood and hardwood, and this is advantageous for pulping. Low lignin content in kenaf was revealed and the success of kenaf has been attributed to the quality of its bast fibers (Villar et al. 2009). Core fibers contain a higher amount of lignin compared to bast fibers, 19.2% and 14.7% lignin, respectively. Lignin is a polyphonic, amorphous, three-dimensional branched network polymer that plays an important part for the mechanical support in plants. Lignin acts as a binder in lignocellulosic plants that hold the fiber together and act as a stiffening agent within the fibers (Biermann 1996).

The values of ash content were 5.4% and 1.9% for bast and core fibers, respectively. The ash content is the inorganic constituent of lignocellulosic materials which is the residue remaining after combustion of organic matter at a temperature of 525 ± 25°C (Rowell et al. 1997). Ash content consists mainly of various mineral salts such as silicates, carbonates, oxalates, and phosphates of potassium, magnesium, calcium, iron, and manganese, as well as silicon. A high ash content is undesirable during refining and recovery of the cooking liquor for pulping (Rodra-Gueza et al. 2008). The ash content of core and bast C fibers are lower than those of non-woods, such as bamboo and rice straw, but higher than wood (Abdul Khalil et al. 2010; H'ng et al. 2009; Ashori 2006).

Compared to wood, both bast and core fibers contain lower extractives, and the extractive content in bast is relatively higher than the core. Extractives are the extraneous plant components that are present in small to moderate amounts, and can be isolated by organic solvents or water. They are heterogeneous groups of compounds of lipophilics and hydrophilics. Generally, the presence of extractives in woody materials increases the consumption of pulp reagents and reduces yield. According to Rodra-Gueza et al. (2008), materials with little or no extractive content are the most desirable.

## 2.6 PHYSICAL PROPERTIES OF KENAF FIBERS

Green kenaf stems have a moisture content (MC) of between 200–400% (Muehl et al. 1999; Juliana et al. 2014). In a study carried out by Juliana et al. (2014), the densities of kenaf whole stem and kenaf core increased from the bottom to the top, which is consistent with the findings obtained by Voulgaridis et al. (2000) due to larger vessel diameters at the bottom compared to the top. The core is a light and porous material, with a density of 0.10–0.20 g/cm³, which is a good material for a low-density board (Kajita et al. 1998; Sellers et al. 1993). Meanwhile, kenaf bast has a density of 1.20–1.50 g/cm³ and is a dense and strong material suitable for composites, automotive interiors, and reinforcement materials (KEFI 2004; Munawar et al. 2007).

To the authors' knowledge, most studies reported on the physical properties of kenaf composites rather than the fiber itself. However, few studies were done on the water absorption of the kenaf fibers (Lips et al. 2009; Nosbi et al. 2010). Kenaf core is an absorbent material especially in the pith part of kenaf (the spongy central part of the core) with the highest water absorption property. Lips et al. (2009) found that the pith material is rather soft, and the absorbed water can easily be squeezed out, but at 940 G it still holds more than five times its own weight. Meanwhile, Nosbi et al. (2010) stated that kenaf bast fibers immersed in sea water for 140 days exhibited the highest water absorption value compared to distilled water and acid solution. The swelling is expected to have close relationship with the chemical composition of kenaf fiber. Theoretically, fibers with higher lignin content would be expected to exhibit the lowest value of water absorption, due to hydrophobic characteristic of lignin thus providing resistance against the hydro-degradation of the kenaf fiber. Therefore, fibers immersed in sea water has lowest lignin content compared to others.

## 2.7 MECHANICAL PROPERTIES OF KENAF FIBERS

The mechanical properties of fibers including the tensile strength and modulus of elasticity usually influences the properties of the final composite. Table 2.3 tabulates the mechanical properties of kenaf and some natural fibers. From Table 2.3, it can be observed that bast fibers (kenaf, hemp, jute, and flax) have higher densities which explains the relatively higher tensile strength compared to the others. Kenaf bast fibers have the greatest tensile strength, ranging from 295 to 1191 MPa, followed by flax, sisal, and hemp. Owing to the superior tensile properties of kenaf bast fibers, it is suitable to be used as reinforcement and for pultruded composites. Based on previous studies, the addition of kenaf fibers to pultruded samples performed better in terms of mechanical and thermal properties, compared to samples that were fabricated through ordinary manufacturing processes (Nosbi et al. 2010; Omar et al. 2010).

A study done by Nurhafizah and Jamaludin (2010) on untreated and treated kenaf bast fibers using 5% solution of sodium hydroxide reported that the average fiber tensile strength and Young's modulus for untreated fibers is 129 MPa and

## TABLE 2.3
### Mechanical Properties of Some Kenaf and Non-Wood Fibers

| Nonwood | Kenaf Bast | Hemp | Jute | Flax | Sisal | Oil Palm Trunk | Oil Palm EFB | Bagasse | Wheat Straw | Coir | Bamboo |
|---|---|---|---|---|---|---|---|---|---|---|---|
| Density $(g/cm^3)$/ Specific gravity | 1.3–1.5[1,2] | 1.48[2] | 1.45[2] | 1.54[2] | 0.76–1.45[1,2] | 0.27–0.44[3] | 0.18–1.32[4] | 0.52–1.47[5] | 0.02–1.10[6] | 1.15[7] | 0.58–0.95[8] |
| Tensile strength (MPa) | 295–1191[1,9] | 310–750[10] | 370[11] | 500–900[12] | 80–840[1,10] | 300–600[13] | 71[14] | 70[15] | 21–28[16] | 106–175[9,17] | 441–575[11] |
| E-modulus (GPa) | 22–60[1,9] | 30–60[10] | 2.5–23[2,11] | 50–70[12] | 9–22[1,10] | 15–32[13] | 1.7[14] | – | – | 2–6[9,17] | 27–36[11] |

*Source:* Paridah, M.T. et al., *Current Forestry Reports*, 1(4), 221–238, 2015.

*Note:*
[1]Munawar et al. 2007
[2]KEFI 2004
[3]Loh et al. 2011
[4]Asiah et al. 2004
[5]Law et al. 2007
[6]Lam et al. 2008
[7]Rao and Rao 2007
[8]Razak et al. 2012
[9]Abdul Khalil 1999
[10]Mwaikambo and Ansell 2006
[11]Okubo and Yuzo 2004
[12]Bolton 1994
[13]Zakiah et al. 2010
[14]Yusoff et al. 2010
[15]Cao et al. 2006
[16]O'Dogherty et al. 1995
[17]Li et al. 2007

12152 MPa, respectively, while the average fiber tensile strength and Young's modulus for treated fibers was 108 MPa and 13281.36 MPa, respectively. This indicated that the treated fibers have lower tensile strength but higher Young's modulus values compared to untreated kenaf fibers. This might be due to the high moisture content after the alkalization process. Once the waxes or substance are removed by sodium hydroxide, the fibers have the tendency to absorb moisture. Even though the treatment significantly improved the Young's modulus of kenaf fibers, kenaf still has a lower Young's modulus compared to Curaua with a value of 64 GPa, which is comparable to that of glass fibers (70 GPa) (Fidelis et al. 2013).

The mechanical properties of plant fibers are influenced by various other properties namely the cellulose content, fiber diameter, microfibrillar angle, and moisture content (Célino et al. 2014). A higher cellulose content indicates better mechanical properties. A high cellulose content, as well as greater fiber length and cell wall thickness, may be responsible for the high strength properties (Horn and Setterholm 1990). As mentioned by Prasad and Sain (2003), tensile properties were found to be clearly dependent on the diameters of the fibers, decreasing gradually with the increase in fiber diameter. This is consistent with the general observation, also applicable to synthetic fibers, that as the fiber diameter decreases, the amount of flaws in the fibers also decreases, thus resulting in an increase in the tensile properties of fibers.

## 2.8 SUMMARY

Kenaf, having a greater yield in comparison to other widely used non-wood fibers such as hemp, flax, and jute is one of the prominent annual bast fibers, and is a promising fiber source for various applications. The two distinct fibers found in the kenaf plant, namely bast and core, present a challenge in their utilization. Due to their vastly different anatomical, chemical, physical, and mechanical properties, separation is necessary for most applications. This will incur additional processing costs and time. However, in many cases, kenaf can still be utilized in its whole stem. The chemical composition of kenaf fibers is relatively similar to those of wood, with the only difference being a higher ash content. Kenaf bast fibers have lengths similar to softwoods while the core fibers are similar to hardwoods. Kenaf bast fibers, with their high density and mechanical strength, have the potential to be utilized in materials requiring high strength, such as composites, automotive applications, and reinforcement. On the other hand, there are challenges to achieve sufficient bonding among kenaf bast fibers, due to the presence of waxy substances on the surface of the fibers. The thick cell walls of these fibers also hinder the penetration of matrices, which can result in poor bonding properties of the final product. Meanwhile, kenaf core fibers have extremely low densities. This makes the fibers suitable in the production of lightweight boards not intended for strength applications. Their low porosity and density also makes them a possible candidate in the manufacture of sound and heat insulation boards.

## REFERENCES

Abdul Khalil, H.P.S. (1999). Acetylated Plant Fibre Reinforced Composites. PhD Thesis. School of Agricultural and Forest Sciences, University of Wales, Bangor, Gwynedd, United Kingdom.

Abdul Khalil, H.P.S., Ireana Yusra, A.F., Bhat, A.H., and Jawaid, M. (2010). Cell Wall Ultrastructure, Anatomy, Lignin Distribution, and Chemical Composition of Malaysian Cultivated Kenaf Fibre. *Industrial Crops and Products* 31(1): 113–121.

Aisyah, H. (2013). (Kenaf. *Hibiscus Cannabinus* L.) Fibres Properties under Varied Thermo-Mechanical Pulping (TMP) Conditions and Their Influence on MDF Performance. Master's Thesis. School of Graduate Studies, Universiti Putra Malaysia, Serdang, Selangor, Malaysia.

Aisyah, H.A., Paridah, M.T., Sahri, M.H., Astimar, A.A., and Anwar, U.M.K. (2013). Properties of Medium Density Fibreboard (MDF) from Kenaf (*Hibiscus cannabinus* L.) Core as Function Of Refining Conditions. *Composite Part B* (44): 592–596.

Ashori, A., Harun, J., Raverty, W.D., and MohdNor, M.Y. (2006). Chemical and Morphological Characteristics of Malaysian Cultivated Kenaf (*Hibiscus cannabinus*) Fiber. *Polymer-Plastics Technology and Engineering* 45(1): 131–134.

Asiah, A., Mohd Razi, I., Mohd Khanif, Y., Marziah, M., and Shaharuddin, M. (2004). Physical And Chemical Properties of Coconut Coir Dust and Oil Palm Empty Fruit Bunch and the Growth of Hybrid Heat Tolerant Cauliflower Plant. *Pertanika Journal Tropical Agricultural Science* 27(2): 121–133.

Azizi Mosello, A., Jalaluddin, H., Seyed Rashid, F.S., Hossein, R., Paridah, M.T., Rushdan, I., and Ainun Zuriyati, M. (2010). A Review of Literatures Related to Kenaf as an Alternative for Pulpwoods. *Agricultural Journal* 5: 131–138.

Biermann, C.J. (1996). *Handbook of Pulping and Papermaking*. 2nd Edition, Academic Press, San Diego, California.

Blackburn, R.S., Editor. (2005). *Biodegradable and Sustainable Fibres*. Cambridge: Woodhead Publishing Series in Textiles: 47, The Textile Institute.

Bolton, A.J. (1994). Natural Fibres for Plastic Reinforcement. *Material Technology* 9: 12–20.

Cao, Y., Shibata, S. and Fukumoto, I. (2006). Mechanical Properties of Biodegradable Composites Reinforced with Bagasse Fibre Before and After Alkali Treatments. *Composite Part A* 37: 423–429.

Célino, A., Fréour, S., Jacquemin, F., and Casari, P. (2013). The hygroscopic behavior of plant fibers: A review. *Frontiers in Chemistry* 1.

Dutt, D., Upadhyay, J.S., Singh, B. and Tyagi, C.H. (2009). Studies on *Hibiscus cannabinus* and *Hibiscus sabdariffa* as an Alternative Pulp Blend for Softwood: An Optimization of Kraft Delignification Process. *Industrial Crops and Products* 29: 16–26.

Farnfield, C.A. (1975). *Textile Terms and Definitions*, 7th Edn. The Textile Institute.

Fidelis, M.E.A., Pereira, T.V.C., Gomes, O.D.F.M., de Andrade Silva, F., and Toledo Filho, R.D. (2013). The Effect of Fiber Morphology on the Tensile Strength of Natural Fibers. *Journal Of Materials Research And Technology* 2(2): 149–157.

H'ng, P.S., Khor, B.N., Tadashi, N., Aini, A.S.N., and Paridah, M.T. (2009). Anatomical Structures and Fibre Morphology of New Kenaf Varieties. *Asian Journal of Scientific Research.* 2: 161–166.

Horn, R.A. and Setterholm, V.C. (1990). Fiber morphology and new crops. In: Janick J., Simon J.E. (Eds.), *Advances in New Crops*. Portland: Timber Press, 270–275.

Juliana, A.H. (2013). Mechanical and Physical Properties of Particleboard Manufactured from Kenaf (*Hibiscus cannabinus* L.) and Rubberwood (*Hevea brasiliensis*). PhD's Thesis. School of Graduate Studies, Universiti Putra Malaysia, Serdang, Selangor, Malaysia.

Juliana, A.H., Paridah, M.T., Rahim, S., Nor Azowa, I., and Anwar, U.M.K. (2012). Production of Particleboard from Kenaf (*Hibiscus cannabinus* L.) as Function of Particle Geometry. *Materials and Design* 34: 406–411.

Juliana, A.H., Paridah, M.T., Rahim, S., Nor Azowa, I., and Anwar, U.M.K. (2014). Effect of Adhesion and Properties of Kenaf (*Hibiscus cannabinus* L.) Stem in Particleboard Performance. *Journal of Adhesion Science and Technology* 28(6): 546–560.

Kajita, H., Kawasaki, T., and Kawai, S. (1998). Proceedings of the 4th Pacific Rim Bio-Based Composites Symposium, Bogor, Indonesia, 479.

Kawai, S. (1999). International Conference on Effective Utilization Of Plantation Timber. Chi-Tou, Taiwan, 109–114.

Kawai, S. (2005). Development of High-Performance Kenaf Bast Oriented Fibreboard and Kenaf Core Binderless Particleboard. *Sustainable Humanospere* 1: 12.

KEFI. (2004). Properties of principal fibers. http://www.kenaf-fiber.com/en/infotec-tabella10.asp. Accessed 10 Dec 2015.

Kozlowski, R. (2000). Potential and Diversified Uses of Green Fibres. Third International Wood and Natural Fibres Composites Symposium, 19th–20th September, Kassel, German, 1–14.

Lam, P.S., Sokhansanj, S., Bi, X., Lim, C.J., Naimi, L.J., Hoque, M., Mani, S., Womac, A.R., Ye, X.P., and Narayan, S. (2008). Bulk Density of Wet and Dry Wheat Straw and Switch Grass Particles. *Applied Engineering Agricultural* 24(3): 351–358.

Law, K.N., Wan Rosli, W.D., and Arniza, G. Morphological and Chemical Nature of Fiber Strands of Oil Palm Empty-Fruit-Bunch (OPEFB) (2007). *Bioresources* 2(3): 351–362

Li, Z.H., Wang, L., and Wang, X. (2007). Cement Composites Reinforced with Surface Modified Coir Fibers. *Journal of Composite Materials* 41(12): 1445–1457.

Lips, S.J., de Heredia, G.M.I., den Kamp, R.G.O., and van Dam, J.E. (2009). Water absorption characteristics of kenaf core to use as animal bedding material. *Industrial Crops and Products* 29(1): 73–79.

Loh, Y.F., Paridah, M.T., and Hoong, Y.B. (2011). Density Distribution of Oil Palm Stem Veneer and Its Influence on Plywood Mechanical Properties. *Journal of Applied Sciences* 11(5): 824–831.

Monti, A. and Alexopoulou, E. (2013). *Kenaf: A Multi-Purpose Crop for Several Industrial Applications*. London, UK: Springer.

Muehl, J.H., Kryzsik, M.A., Youngquist, A.J., Chow, P., and Bao, Z. (1999). Performance of Hardboard from Kenaf, In. Terry Sellers, J.R., Nancy A. Reichert, Eugene P. Columbus. Marty J. Fuller, and Karen Williams (Eds.), *Kenaf Properties: Processing and Products*. Mississippi State University.

Munawar, S.S., Umemura, K. and Kawai, S. (2007). Characterization of the Morphological, Physical and Mechanical Properties of Seven Nonwood Plant Fiber Bundles. *Journal of Wood Sciences* 53: 108–113.

Mwaikambo, L. (2006). Review of the History, Properties and Application of Plant Fibres. *African Journal of Science and Technology* 7(2): 121.

Mwaikambo, L.Y. and Ansell, M.P. (2006). Mechanical Properties of Alkali Treated Plant Fibres and Their Potential as Reinforcement Materials.1. Hemp Fibres. *Journal of Material Sciences* 41: 2483–2496.

Nosbi, N., Akil, H.M., MohdIshak, Z.A., and Abu Bakar, A. (2010). Degradation of Compressive Properties of Pultruded Kenaf fibre Reinforced Composites after Immersion in Various Solutions. *Material Design* 31: 4960–4964.

Nur Hafizah, A.K. and Jamaludin, M.Y. (2010). Tensile Behavior of the Treated and Untreated Kenaf Fibres. In 2010 National Postgraduate Seminar (NAPAS 10′) Kuala Lumpur, Malaysia. Bridging Postgraduate Research Towards Industry Linkage and Future Innovation July 6th–7th 2010.

O'Dogherty, M.J., Huber, J.A., Dyson, J., and Marshall, C.J. (1995). A Study of the Physical and Mechanical Properties of Wheat Straw. *Journal of Agriculture Engineering Research* 62(2): 133–142.

Ohnishi, K., Umeoka, K., Okudaira, Y., Zhang, M., and Kawai, S. (2003). Development of Kenaf Boards and Their Properties. Proceedings of the International Kenaf Symposium, CCG International, Beijing, China, 19–21 August, 175–188.

Okubo, K. and Yuzo, T.F. (2004). Development of Bamboo-Based Polymer Composites and Their Mechanical Properties. *Composite Part A* 35: 377–383.

Omar, M.F., Md Akil, H., Ahmad, Z.A., Mazuki, A.A.M., and Yokoyama, T. (2010). Dynamic Properties of Pultruded Natural Fibre Reinforced Composites Using Split Hopkinson Pressure Bar Technique. *Material Design* 31: 4209–4218.

Pande, H. and Roy, D.N. (1998). Influence of Fibre Morphology and Chemical Composition on the Papermaking Potential of Kenaf Fibres: A Look at What Attributes Affect Tensile Strength. *Pulp & Paper Canada* 99(11): 31–34.

Paridah, M.T., Nor Hafizah, A.W., Zaidon, A., Azmi, I., Mohd Nor, M.Y., and Nor Yuziah, M.Y. (2009a). Bonding Properties and Performance of Multi-Layered Kenaf Board. *Journal of Tropical Forest Science* 21(2): 113–122.

Paridah, M.T. and Khalina, A. (2009b). Effects of Soda Retting on The Tensile Strength of Kenaf (*Hibiscus cannabinus* L.) Bast Fibres, Project Report Kenaf EPU, 21.

Paridah, M.T., Juliana, A.H., Zaidon, A., and Khalil, H.A. (2015). Nonwood-Based Composites. *Current Forestry Reports* 1(4): 221–238.

Prasad, B.M. and Sain, M.M. (2003). Mechanical Properties of Thermally Treated Hemp Fibers in Inert Atmosphere for Potential Composite Reinforcement. *Materials Research Innovations* 7(4): 231–238.

Rao, K.M.M. and Rao, K.M. (2007). Extraction and Tensile Properties of Natural Fibers: Vakka, Date and Bamboo. *Composite Structures* 77: 288–295.

Razak, W., Mohd Tamizi, M., Shafiqur, R., Mohammed, A.S., Othman, S., Mahmud, S. and Mohd Sukhairi, M.R. (2012). Relationship between Physical, Anatomical and Strength Properties of 3-Year-Old Cultivated Tropical Bamboo *Gigantochloascortechinii*. *ARPN Journal of Agriculture and Biological Sciences* 7(10): 782–791.

Rodra-Gueza, A., Morala, A., Seraanoa, Labidib, J., and Jimanez, L. (2008). Rice Straw Pulp Obtained by Using Various Methods. *Bioresource Technology*. 99: 2881–2886.

Rowell, R.M. and Stout, H.P. (1998). Jute and kenaf. In: *Handbook of Fiber Chemistry*; Lewin, M. and Pearce, E.M. (Eds.), Marcel Dekker, Inc, New York, 466–502.

Rowell, R.M., Young, R.A., and Rowell, J.K. (1997). *Paper and Composites from Agro-based Resources*. Lewis, New York.

Rymsza, T.A. (2001). Commercial Paper Making With kenaf. Paper presented at the PIRA Nonwoods Conference, Amesterdam, NL, October.

Sellers, T. Jr., Miller, G.D., and Fuller, M.J. (1993). KenafCore as a Board Raw Material. *Forest Products Journal* 43: 69–71.

Sen, H.S. (2009). Quality improvement in jute and kenaf fibre. Proceedings of The International Conference on Prospects of Jute & Kenaf as Natural Fibres, International Jute Study Group, Dhaka, Bangladesh.

Smole, M.S., Hribernik, S., Kleinschek, K.S., and Kreže, T. (2013). Chapter 15: Plant Fibres for Textile and Technical Applications. *Advances in Agrophy Research* 1: 52372.

Thi Bach, T.L., Keko, H., and Kenji, I. (2003). Structural Characteristics of Cell Walls of Kenaf (*Hibiscus cannabinus* L.) and Fixation of Carbon Dioxide. *Journal of Wood Science* 49: 255–261.

Touzinsky, G.F. (1987). Kenaf. *Pulp and Paper Manufacturing*, Vol. 3, Secondary Fibers and Non-Wood Pulping, Chapter 8. TAPPI Press, Atlanta, GA 106.

Tsoumis, G. (1991). *Science and Technology of Wood: Structure, Properties and Utilization.* Van Nostrand Reinhold, New York.

Ververis, C., Georghiou, K., Christodoulakis, N., Santas, P., and Santas, R. (2004). Fiber dimensions, lignin and cellulose content of various plant materials and their suitability for paper production. *Industrial Crops and Products* 19(3): 245–254.

Viilar, J.C., Revilla, E., Gomez, N., Carbajo, J.M., and Simon, J.L. (2009). Improving the Use of Kenaf for Kraft Pulping By Using Mixtures of Bast and Core Fibres. *Industrial Crops and Products.* 29: 301–307.

Vincent, J.F. (2000). A Unified Nomenclature for Plant Fibres for Industrial Use. *Applied Composite Materials* 7(5–6): 269–271.

Voulgaridis, E., Passialis, C., and Grigoriou, A. (2000). Anatomical Characteristics and Properties of Kenaf Stem (*Hibiscus cannabinus*). *IAWA Journal* 21(4): 435–442.

Westman, M.P., Laddha, S.G., Fifield, L.S., Kafentzis, T.A., and Simmons, K.L. (2010). *Natural Fiber Composites: A Review.* Pacific Northwest National Laboratory.

Wood, J.R. (2000). Wood-induced Variations in TMP Quality-Their Origins and Control. TAPPI Pulping/pocess and product quality conference, Proceedings 491–497.

Yusoff, M.Z.M., Sapuan, M.S., and NapsiahI, R.W. (2010). Mechanical Properties of Short Random Oil Palm Fibre Reinforced Epoxy Composites. *Sains Malaysiana* 39(1): 87–92.

Zakiah, A., Hamami, S., and Paridah, M.T. (2010). Oil Palm Trunk Fiber as a Bio-Waste Resource for Concrete Reinforcement. *International Journal Mechanical Materials Engineering* 5(2): 199–207.

# 3 Adhesion Characteristics of Kenaf Fibers

*M.T. Paridah and A.H. Juliana*

## CONTENTS

3.1  Introduction ...........................................................................................................37
3.2  Adhesion Principle ................................................................................................38
3.3  Bonding Mechanisms in Natural Fibers ...............................................................41
3.4  Kenaf Fibers ..........................................................................................................42
    3.4.1  Fiber Morphology of Kenaf ....................................................................42
    3.4.2  Adhesion Properties of Kenaf .................................................................43
    3.4.3  Surface Energy and Contact Angle .........................................................44
    3.4.4  Wetting and Wettability ..........................................................................46
    3.4.5  Buffering Capacity ..................................................................................47
    3.4.6  Wettability of Kenaf ................................................................................49
    3.4.7  Buffering Capacity of Kenaf ...................................................................50
    3.4.8  Bonding Limitation in Kenaf Fibers .......................................................53
    3.4.9  Surface Inactivation and Adhesive Bonding ..........................................54
    3.4.10 Surface Inactivation ................................................................................54
3.5  The Compatibility of Kenaf Fibers and Polymers ...............................................55
3.6  Future Challenges .................................................................................................55
References .....................................................................................................................58

## 3.1 INTRODUCTION

Adhesion mechanisms are dependent on the surface characteristics of various materials. Since composites were industrialized a century ago, there is tremendous increase in the understanding of adhesion mechanisms as a result of aggressive efforts by both the automotive and aerospace industries in seeking lighter and cheaper alternatives to metals and metal components. The rise of natural fiber-reinforced composites since the end of the last century has also contributed to the advancement in adhesion mechanism studies. In principle, there are three components for adhesion to occur viz. substrates, adhesive matrices, and interfaces. Traditionally, fiber-reinforced composites use substrates such as glass; carbon; aluminium oxide; and plastic polymers matrices, including high density polyethylene (HDPE), low density polyethylene (LDPE), polypropylene (PP), polyether ether ketone (PEEK). However, for the past several decades, there is a significant growth in the use of natural fibers, especially bast (bark) fibers (such as kenaf, flax, hemp, jute, henequen) and other types of matrices like phenolic, epoxy, polylactic acid (PLA), modified starch, etc. This drive has been the major influence in the need to understand polymer adhesion and to resolve the debate over how the interfaces are actually adhering (Awaja et al. 2009). Being porous, the bond formation in natural fibers generally involves three main adhesion mechanisms: mechanical interlocking, adsorption theory, and chemical bonding. The following sections give a brief overview of adhesion mechanisms in relation to bonding of kenaf fibers.

## 3.2 ADHESION PRINCIPLE

Adhesion is the state that in which two surfaces are held together by interfacial forces which consist of valence forces and/or interlocking actions. These forces of attraction are caused by the presence of interactions between molecules, atoms, or ions on the two surfaces involved in the adhesion process. Three major components in gluing are substrate, adhesive, and interface. There are several theories that attempt to explain the phenomenon of adhesion of the adhesive/matrix on the substrates, however, there is currently no unified theory to justify all cases. Nevertheless, Packham (2011) listed four common theories to describe this phenomenon: mechanical interlocking, adsorption, electrostatic, and diffusion. In most cases, these theories are combined to describe the actual bonding process in natural fibers.

*Mechanical Interlocking*

The mechanical interlocking adhesion mechanism is based on the adhesive deposited in the voids of any porous substrate that, once cured, form an anchor between the adhesive and the substrate. Any type of material when observed at the microscopic level has a surface composed of valleys and ridges. This surface topography allows the adhesive to penetrate and fill the valleys, resulting in anchorage areas between the adhesive and substrate. The mechanical interlocking begins with the adhesive first being spread onto the surface of the porous substrate (Figure 3.1), forming interface layers between the adhesive and the substrates. Because the substrates are porous, the adhesive (normally in liquid form) flows through the cell wall and into the lumen while at the same time forming several layers: boundaries between interface and substrate, between two interfaces, and between interface and adhesive. These layers depend upon each other like a chain; if one layer breaks, the whole bond collapses. Such an incidence can be represented by a nine-chain concept (Marra 1992).

Apart from the roughness and porosity of the substrate surface, it is necessary that the adhesive/matrix has a good filling power (i.e., suitable viscosity), and can penetrate into the valleys and pore substrate surface to generate adhesion anchor points. Mechanical adhesion theory does not account for the incompatibility that may exist between the adhesive and the substrate, it only

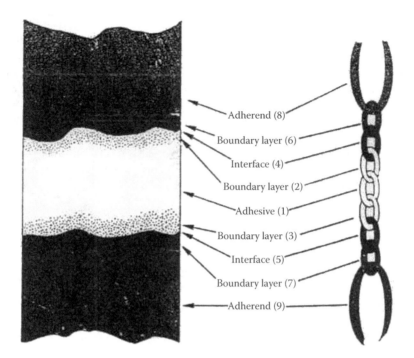

**FIGURE 3.1** Illustration of mechanical interlocking between two substrates.

takes into account the topography of the substrate and adhesive filling power. Therefore, this theory cannot explain the adhesion between surfaces with low roughness or smooth, nor the lack of adhesion between rough substrates incompatible with adhesives.

*Adsorption Theory*

The adsorption theory or model explains the phenomenon of adhesion based on concepts such as contact angle, wet ability (also denoted as wettability) and surface tension. When the adhesive has a lower surface tension compared to the substrate surface energy, it is capable of wetting the surface, generating a contact angle less than 90°, thus generating the adhesion between the adhesive and substrate. Against the mechanical model and the model of diffusion, the adsorption model explains the phenomenon of adhesion without penetration by the adhesive to the substrate; the adhesion is generated by the contact between the adhesive and substrate.

*Chemisorption Theory*

Chemisorption theory is an extension of the adsorption theory of adhesion, in which the adhesive has properly wet the substrate; the adhesion phenomenon arises when generating intermolecular or van der Waals forces and chemical bonds between the adhesive and substrate. This theory is very much suited to elucidate the use of compatible agents in fiber-reinforced composites.

*Diffusion Theory*

The diffusion model explains the concept of adhesion by the compatibility between polymers and the movements that occur in the polymer chains. When two polymers are compatible, their polymer chains are able to mix between them, resulting in partial penetration between the two materials; as a result of these penetrations, anchorage areas and adhesion points are formed. The mobility and degree of penetration of the polymers are determined directly by their molecular weight, so that short polymer chains have high mobility and penetrate into the other material before long polymer chains. This theory can explain the phenomenon of adhesion that occurs between polymeric materials, plastic welding, and plastic binding with adhesives.

*Chemical Bonding*

Chemical bonding is the most widely accepted mechanism for explaining adhesion between two surfaces in close contact. It entails intermolecular forces between adhesive and substrate such as dipole-dipole interactions, van der Waals forces, and chemical interactions (that is, ionic, covalent, and metallic bonding). This mechanism describes the strength of the adhesive joints by interfacial forces and also by the presence of polar groups (Awaja et al. 2009). Chemical bonding mechanisms require an intimate contact between the two substrates, as shown in Figure 3.2. Among the types or classes of chemical bonds mentioned above, the covalent bond is the one that takes place in the adhesive joints using organic adhesives based on polymers.

Intimate contact alone is often insufficient for good adhesion at the interface due to the presence of defects, cracks, and air bubbles. The molecular bonding mechanism is not yet fully understood and there have been many theories proposed to explain it. Chen et al. (2007), who investigated the shear strength of aluminium–polypropylene lap joints, found that the overriding adhesive mechanism was the chemical interaction between the functional groups at the interface and also concluded that excessive

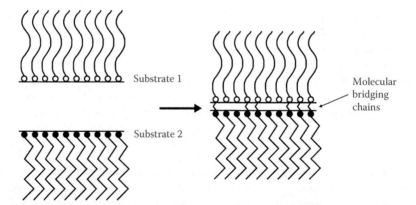

**FIGURE 3.2** Schematic of the molecular bonding between substrates. (From Awaja, F. et al, *Progress in Polymer Science*, 34: 948–968, 2009.)

chemical bonding at the adhesive interface could have a negative effect on the interface strength. Such conditions may occur as the number of chemical bonds increase at the contact zone, adhesion strength passes through a maximum value. Once the number of chemical bonds increases past this value, the concentration of the mechanical stresses at the interface leads to a decrease in adhesion strength due to the excessive increases in the size of the interfacial effect. The fact that adhesion strength depends also on the thickness of the adhesive layer for composite interfaces renders such an explanation valid. Interfacial bonding strength increases as the thickness of the adhesive layer is reduced (i.e., a thinner glue line) as stress is able to dissipate through the interface with greater ease.

## 3.3 BONDING MECHANISMS IN NATURAL FIBERS

A strong fiber–matrix interface bond is critical for high mechanical properties in composites. In polymer matrix composites, there appears to be an optimum level of fiber-matrix adhesion, which provides the best composite mechanical properties (Drzal et al., 2005). Physical properties of natural fibers are basically influenced by the chemical structure such as cellulose content, degree of polymerization, orientation, and crystallinity, which are affected by conditions during growth of the plants as well as extraction methods used. There is an enormous amount of variability in fiber properties depending upon from which part of the plant the fiber is taken, the quality of the plant, and the plant's location. Different fibers have different lengths and cross-sectional areas and also different defects such as micro compressions, pits, or cracks. To obtain uniformity and flexibility, some researchers conduct surface modification on the fibers by dissolving the microfibrils in solvents followed by precipitation under controlled conditions, or by increasing the wettability of the fiber surfaces through pretreatment with chemicals, or by adding coupling agents.

The natural fiber-polymer interface presents a formidable challenge. Due to the presence of hydroxyl and other polar groups in various constituents of natural fibers, the moisture absorption in biocomposites is high, which leads to weak interfacial bonding between fibers and matrix polymers. Natural fibers have a strong polar character that renders hydrophilic properties, whilst thermosetting and thermoplastic matrices are hydrophobic in nature. Hence, when both are combined, it is necessary to make use of compatibilizers or coupling agents in order to improve the adhesion between fiber and matrix (Bledzki and Gassan 1999). As many other fiber crops, the bonding of kenaf is relatively more difficult mainly due to lower wettability. Drzal and co-workers (2005) made a comprehensive review of the influence of various surface modifications of natural fibers like henequen, jute, and coconut (coir) fibers and their effects on performance of biocomposites. According to them, the main drawback of natural fibers is their hydrophilic nature, which lowers their compatibility with hydrophobic polymer matrices. The hydrophilic nature of biofibers leads to biocomposites with higher water absorption characteristics that reduce their usefulness in many applications. The presence of a natural waxy substance on the fibers' surface contributes immensely to ineffective fibers in polymer matrix bonding and poor surface wetting. The presence of free water and hydroxyl groups, especially in the amorphous regions, reduces the ability of natural fibers to develop acceptable adhesive characteristics with most binder resins. High water and moisture absorption of the fibers causes swelling and a plasticizing effect, resulting in dimensional instability and poor mechanical properties.

## 3.4 KENAF FIBERS

Most of our work has been on the use of kenaf in particleboard, medium density boards (MDF), fiber-reinforced composites (FRC), and pulp and paper and woven composites (Fariborz et al. 2016; Moradbak et al. 2015; Nayeri et al. 2013; Juliana et al. 2012; Aisyah et al. 2012; Ahmed et al. 2013; Azizi Mossello et al. 2010; Juhaida et al. 2009). We have also extensively studied the retting process of kenaf stems using water, chemicals, and microbes for the production of long fibers (Paridah et al. 2011). Generally, kenaf bast fiber bundles are composed of elongated, thick-walled ultimate cells that are joined together both end-to-end and side-by-side, forming aggregates of fiber bundles along the height of the plant stem. During the growing period of the stem, a circumferential layer of primary fibers are developed from the protophloem, but, as vertical growth ceases in the lower parts, the secondary phloem fibers (where the bast fibers can be obtained) are developed as a result of cambial activity.

Depending on the location in the stem, kenaf contains three types of fiber: bast, core, and pith. The fibers from the bast are long and have thick cell walls, while the core fibers are thinner and shorter in length (Paridah et al. 2008). The core fibers appear as wedge-shaped bundles of cells intermingled with parenchyma cells and other soft tissue. The pith consists exclusively of parenchymatous cells, which are not typically prismatic but polygonal in shape. In mature plants, kenaf can reach a height of 2.5–3.5 m (Rowell and Stout 2006). Zhang (2003) reported that the kenaf fibers are shorter at the bottom of the stalk and longer at the top. The increase in length from the bottom to the top was found not to be gradual, but S-shaped (Rowell and Han 1999). It was reported that the fiber length increases during the early part of plant growth, and decreases again as the plants mature (Chen et al. 1995). Kenaf single fibers are only about 1–7 mm long and about 10–30 microns wide, thus too short for textile processing (Calamari 1997). Compared with cotton fiber, these fibers are coarse, brittle, and non-uniform, which makes them difficult to be processed using conventional textile or nonwoven fabric equipment. Table 3.1 compares the morphology of kenaf with other types of bast plants.

### 3.4.1 FIBER MORPHOLOGY OF KENAF

Fiber extraction procedures depend on the type of plant and portion thereof from which the fibers are derived (e.g., bast, leaves, and wood), as well as the required fiber performance and economics. Fiber-bearing plants have very different anatomies (e.g., trees versus herbaceous plants) and often fibers are derived from agricultural residues or by-products from industry. Consequently, their processing needs can differ greatly. Wood is primarily composed of hollow, elongated, spindle-shaped cells (called tracheids or fibers) that are arranged parallel to each other along the trunk of the tree. These fibers are firmly cemented together and form the structural component of wood tissue. Fibers are extracted from the wood either mechanically or chemically during the retting or pulping process.

**TABLE 3.1**
**Morphology of Natural Cellulosic Fibers**

| Type of Fiber | Cell Type | Cross-Sectional Shape of Ultimate Cell |
|---|---|---|
| Jute<br>Mesta<br>Kenaf | Multicellular | Polygonal with slightly rounded corners and medium-sized lumen |
| Ramie | Multicellular | Elongated ellipse with collapsed elongated lumen |
| Flax | Multicellular | Appreciable roundness in the corners and medium size lumen |
| Hemp<br>Pineapple | Multicellular | Oval cross-section with collapsed small size lumen |
| Sisal | Multicellular | Polygonal with sharp corners and medium to large size lumen |
| Coir | Multicellular | Polygonal with rounded corners and large size lumen |
| Cotton | Unicellular | Peanut-shaped cross-section of each fiber with elongated collapsed lumen |

*Source:* Sur, D., *Understanding Jute Yarn*, Institute of Jute Technology, Anindita Sur, Kolkata, p. 254, 2005.

Kenaf, like flax, is a bast fiber, hence it has considerably different structures compared to other woody materials and consequently is processed quite differently. Kenaf bast fiber is found in the inner bark of the stems and typically account for less than 30% of the stem weight. Inside the inner bark is a woody core (called the "shive") with much shorter fibers. Fiber strands are removed from the bast via a natural retting or mechanical process. Using natural retting, these fiber strands are several meters long and are actually fiber bundles of overlapping single ultimate fibers. Table 3.2 compares the morphology of kenaf fibers with those from other types of natural fibers.

### 3.4.2 ADHESION PROPERTIES OF KENAF

Adhesive bonding is being used in biocomposites for various applications like construction, transportation, aeronautics, sports, and even repairs. Because all plant fibers are made up of layers of naturally occurring cellulose-hemicellulose-lignin polymers, the variations on the fiber surfaces are inevitable, thus preparation of natural fibers is a critical step. Good bonds are produced but questions remain: What are the appropriate techniques to inspect surfaces? What are the key factors deciding a good or poor bond? How to predict material and surface preparation compatibility? Investigations into the effects of various surface preparation procedures and material systems on the adherent surface chemistry or structure and its relation to subsequent bond performance can be performed using various techniques such as surface appearance or topography, surface chemistry, surface energy or wetting, and fracture evaluation. Wetting and fracture are most easily adapted to laboratory assessment, factory inspection, and repair settings. The following sections review surface tension theory and its relation to surface wetting. Results on wetting and buffering capacity of kenaf are compared with those of oil palm fibers and bamboo.

## TABLE 3.2
### Dimensions of Different Natural Fibers

| Parts Where Fibers Are Obtained | Property (mm) | | Types of Fiber | | |
|---|---|---|---|---|---|
| | | **FLAX** | **HEMP** | **JUTE** | **KENAF** |
| Bast | Length | 4–69 | 5–55 | 0.7–6 | 2–11 |
| | Diameter | 8–31 | 16 | 15–25 | 13–33 |
| Core | Length | 0.2 | 0.7 | 1.06 | 0.6 |
| | Diameter | – | – | 26 | 30 |
| | | **ABACA** | **SISAL** | **PALM** | |
| Leaf | Length | 2–12 | 0.8–7 | 1.2–2.5[3] | |
| | Diameter | 6–40 | 8–48 | 18 | |
| | | **COIR** | **COTTON** | | |
| Seed | Length | 0.2–1 | 10–50 | | |
| | Diameter | 6.24 | 12–25 | | |
| | | **CORN** | **RICE** | **WHEAT** | |
| Straws/stalk | Length | 1.0–1.5 | 0.65–3.48 | 1.5 | |
| (Cereal fibers) | Diameter | 20 | 5–14 | 15 | |
| | | **BAGASSE** | **BAMBOO** | | |
| Stem | Length | 2.8 | 2.7–4 | | |
| (Grass fibers) | Diameter | 34.1 | 15 | | |
| | | **RUBBER-WOOD** | **RADIATA PINE** | | |
| Wood[2] | Length | 1.15–1.34 | 2.6–3.7 | | |
| | Diameter | 0.12–0.28 | 0.37–0.43 | | |

*Sources:* [1]Jusoh, M.J., Studies on the properties of woven natural fibers reinforced unsaturated polyester composites, MSc thesis, Universiti Sains Malaysia, Penang, Malaysia, 2008; [2]Naji, H.R. et al., *BioResources Journal*, 7(1), 189–202, 2012; [3]Ghosh, S.K. et al., *Journal of Agricultural Engineering*, 43(4), 89–91, 2006

### 3.4.3 Surface Energy and Contact Angle

This theory considers the interaction of surface energy of the substrate and the adhesive in relation to bond quality (Sharfrin and Zisman 1960). Surface energy is normally used to indicate the ease of bonding of a surface. This theory involves Young's equation, where θ is an angle between the horizontal plane of a substrate and the contact line of a droplet is denoted as contact angle (Figure 3.3). A contact angle is defined as the angle formed by a drop of liquid adhesive in contact with the surface of

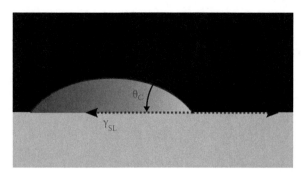

**FIGURE 3.3** Schematic illustration of a liquid drop showing the quantities in Young's equation.

the substrate; the value of this angle will indicate the degree of wettability and therefore the adhesion of the adhesive on the substrate.

Adhesive forces between a liquid and solid cause a liquid drop to spread across the adhesive forces between the surfaces. Cohesive forces within the liquid cause the drop to ball up and avoid contact with the surface. The contact angle (θ), as seen in Figure 3.4, is the angle at which the liquid–vapor interface meets the solid–liquid interface. The contact angle is determined by the resultant adhesive and cohesive forces. As the tendency of a drop to spread out and increase over a flat, solid surface, the contact angle decreases. Thus, the contact angle provides an inverse measure of wettability.

A contact angle of less than 90° (low contact angle) usually indicates that wetting of the surface is very favorable, and the fluid will spread over a large area of the surface. At this state, the adhesive wets the substrate causing adhesion between both materials. Complete wetting means θ approaches zero. On the other hand, contact angles greater than 90° (high contact angle) generally means that wetting of the surface is unfavorable, so the fluid will minimize contact with the surface and form a compact liquid droplet. Under this condition, it is said that the adhesive does not wet the substrate—it does not generate any adhesion between the adhesive and the substrate. Contaminants usually lower the solid's surface energy (increase θ). The contact angle of an adhesive on any substrate can be determined by using specific devices, such as goniometers or by applying specific software.

 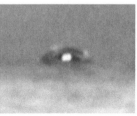 

**FIGURE 3.4** Wetting of kenaf core surface: A water droplet after 30 sec (left); 1 min (center); and 5 min (right). (From Paridah, M.T. et al., *Journal of Tropical Forest Science* 21(2):113–122, 2009.)

## TABLE 3.3
### Relationships between Contact Angle and the Degree of Wetting

|  |  | Strength Of: | |
| --- | --- | --- | --- |
| Contact Angle | Degree of Wetting | Solid/Liquid Interactions | Liquid/Liquid Interactions |
| θ = 0 | Perfect wetting | Strong | Weak |
| 0 < θ < 90 | High wettability | Strong | Strong |
|  |  | Weak | Weak |
| 90° ≤ θ < 180° | Low wettability | Weak | Strong |
| θ = 180° | Perfectly non-wetting | Weak | Strong |

*Source:* Eustathopoulos, N. et al., *Wettability at High Temperatures*. Oxford, UK: Pergamon, ISBN 0-08-042146-6.

For water, a wettable surface may also be termed hydrophilic and a non-wettable surface hydrophobic. Super hydrophobic surfaces have contact angles greater than 150°, showing almost no contact between the liquid drop and the surface. This is sometimes referred to as the "lotus effect." Table 3.3 describes varying contact angles and their corresponding solid/liquid and liquid/liquid interactions (Eustathopoulos et al. 1999). For non-water liquids, the term lyophilic is used for low contact angle conditions and lyophobic is used when higher contact angles result. Similarly, the terms omniphobic and omniphilic apply to both polar and polar liquids.

### 3.4.4 Wetting and Wettability

Wetting is the ability of a liquid to maintain contact with a solid surface, resulting from intermolecular interactions when the two are brought together. Intermolecular forces are defined as the set of attractive and repulsive forces that occur between the molecules as a result of the polarity of the molecules. Intermolecular forces acting between the molecules are classified as permanent dipoles, induced dipoles, dispersed dipoles, and hydrogen bonds.

Within these four groups, the most important forces are the first three, also known as van der Waals forces. The binding energies generated by the intermolecular forces are smaller than the energies generated in the chemical bonds, but exist in greater numbers compared with the number of chemical bonds, so overall they play a very important role. Compared to covalent bonds, the binding energies of van der Waals are 0.1–10 kJ/mol, compared to those of covalent bonds, which are 250–400 kJ/mol. Table 3.4 compares intermolecular forces and chemical bonds.

Wetting is important in the bonding or adherence of two materials. The degree of wetting is determined by a force-balance between adhesive and cohesive forces. Wetting deals with the three phases of materials: gas, liquid, and solid. It has become a topic of interest in nanotechnology and nanoscience studies due to the advent of many nanomaterials in the past two decades (e.g., graphene and carbon nanotubes). Wetting and the surface forces that control wetting are also responsible for other

## TABLE 3.4
## Comparison between Intermolecular Forces and Chemical Bonds

| Intermolecular Forces | Chemical Bonds |
|---|---|
| 1. Intermolecular forces depend on the temperature; an increase of the temperature produces a decrease of the intermolecular forces. | 1. Chemical bonds do not depend so much on the temperature. |
| 2. They are weaker than chemical bonds, to the order of 100 times lower. | 2. They are stronger than intermolecular forces. |
| 3. The bond distance is at the level of microns. | 3. The bond distance is very small, in terms of Angstroms. |
| 4. Unions are not directed. | 4. Unions are directed. |

related effects, including the so-called capillary effects. Regardless of the amount of wetting, the shape of a liquid drop on a rigid surface is roughly a truncated sphere. Various degrees of wetting can be obtained depending on the surface tension of the liquid and the substrate. For instance, both the bast and core of kenaf have a different degree of wetting. As shown in Figure 3.4, kenaf core was completely wet after 5 min (Paridah et al. 2009). Kenaf bast, however, remained unchanged, resembling the droplet on the left even after 20 min.

Wetting is not always sufficient for a strong bond, and that contact angle has been shown to detect changes in surfaces due to contamination or chemical modification. Wettability is one of the indicators of how well the substrate reacts with liquid. It is a quick method for predicting the gluability of unknown materials. Wetting of the surface by an adhesive is a necessary prerequisite to bond formation. A convenient and fast method to measure wetting of a solid surface is through the determination of contact angle of a liquid.

### 3.4.5 Buffering Capacity

Another property that has great influence on the glueline formation is pH and buffer capacity. Buffer capacity measures the resistance of wood to change its pH level either in acids or in alkalines (Paridah et al. 2011). This characteristic is very important for the bonding of porous materials like plant fibers, as it influences the pH of the adhesive/matrix at the boundary layers which eventually affects the rate of curing of the adhesive/matrix.

In conventional biocomposite manufacturing, thermosetting resin, namely urea formaldehyde (UF), melamine urea formaldehyde (MUF), and phenol formaldehyde (PF), are commonly use as binders. These adhesives are sensitive to the pH of the substrate since the rate of cross-linking of most thermosetting adhesives is pH-dependent (Blomquist et al. 1981). Therefore, the formulation of most adhesives is adapted to the acid range and buffer capacity of the substrate. The effect of wood pH and buffering capacity has also been extensively studied in wood composite manufacture because of its effect on the curing of resins. For example, UF resin can decrease the gelation time when in contact with excessively acidic wood

## TABLE 3.5
## pH, Wettability and Buffering Capacity of Kenaf in Comparison to Other Natural Fibers

| Natural Fibers | Treatment/Part | pH | Initial Contact Angle (°) | Buffering Capacity (HCl) | Buffering Capacity (NaOH) |
|---|---|---|---|---|---|
| Rubberwood[1] | Untreated | 5.49 | 58.6 | 10.5 | 6.4 |
| Kenaf[2] | Whole stem | 6.55 | – | 15.5 | – |
| | Core | 6.82 | – | 7.4 | 11.7 |
| | Outer core | – | 52.0 | – | 2.4 |
| | Inner core | – | 44.5 | – | – |
| | Bast | 7.94 | – | 24.8 | 5.9 |
| | Outer bast | – | 68.0 | – | – |
| | Inner bast | – | 60.7 | – | – |
| Empty Fruit Bunch (EFB)[3] | Untreated | 4.26 | – | 2.4 | 8.8 |
| | NaOH soaking | 7.80 | – | 13.2 | 8.4 |
| | Water boiling | 4.53 | – | 1.6 | 12.4 |
| | NaOH soaking and boiling | 7.07 | – | 6.6 | 11.8 |
| Oil Palm Stem (OPS)[4] | Untreated | 5.6 | – | – | – |
| | Outer | | 47 | 1.5 | 9 |
| | Inner | | 46 | 3 | 10 |
| | LMwPF | 6.5 | – | – | – |
| | Outer | – | 79 | 7 | 6 |
| | Inner | – | 92 | 7 | 8 |
| | MMwPF | 6.6 | – | – | – |
| | Outer | – | 62 | 9 | 6 |
| | Inner | – | 67 | 7 | 5 |
| | Commercial PF | 7.8 | – | – | – |
| | Outer | – | 50 | 11 | 5 |
| | Inner | – | 52 | 12 | 5 |
| Bamboo[5] (*Gigantochloascortechinii*) | Untreated | 6.42 | – | 7.5 | 1.0 |
| | Outer | – | 15.3 | – | – |
| | Inner | – | 14.6 | – | – |
| | PF-treated | 5.87 | – | 7.4 | 1.1 |
| | Outer | – | 42.5 | – | – |
| | Inner | – | 42.5 | – | – |

Sources: [1]Paridah Md. Tahir et al., *BioResources* 6(4): 5260–5281, 2011; [2]Juliana, A.H. et al., *J Mater Des* 34:406–411, 2012; [3]Norul Izani, M.A. et al., *Composites B*, 45, 1251–1257, 2013; [4]Wahab, N.H.A. et al., *BioResources*, 7(4), 4545–4562, 2012; [5]Anwar, U.M.K. et al., *J. Bamboo Rattan* 5(3&4):127–133, 2006.

(Freeman and Wangaard 1960). Similarly, the curing of PF resins is negatively affected by wood because of a high alkaline buffering capacity. It was reported that the pH and acid buffering capacity of aqueous extracts from crop materials were significantly higher than those of softwoods, and in the presence of such materials, resin gel time increased greatly.

While contact angle indicates the ease of a liquid to spread and wet the surface, buffering capacity is a measure of resistance of wood or fiber to change in its pH level, either acidic or alkaline. Both the pH and buffering capacity significantly influence the rate of curing of the resin particularly at the interface and the boundary layers. In our study, the bonding characteristics of kenaf was evaluated by determining the wettability through contact angle measurement and the buffering capacity. Table 3.5 presents results of wettability and buffering capacity of different parts of kenaf (outer core, inner core, outer bast, and inner bast) in comparison with other plant fibers.

### 3.4.6 Wettability of Kenaf

Figures 3.5 and 3.6 show that contact angle of wettability reduces as a function of time; a high contact angle was obtained at the initial liquid droplet as measured in 0.1 N HCl and 0.1 N NaOH. The trend is the same for all types and origins of fibers, such as rubberwood, outer core (KOC), inner core (KIC), outer bast (KOB), and inner bast (KIB). Initially, KOB has the highest contact angle, followed by KIB, RW, KOC, and KIC. KIC has a low contact angle, revealing that it has high wettability, as evidently shown by spontaneous wetting which happens within 1 min. This is due to the low specific gravity of the kenaf core plus the large number of voids in it. Meanwhile, RW took about 8 min for a complete wetting. It is obvious that kenaf fibers obtained from the bast have different bonding properties than those obtained from other parts of the stem. Fibers from this part appear to be low in wetting thus some manipulation in resin viscosity and/or incorporation of a coupling agent may be required to help adhesive/matrix penetration.

The fibers behaved differently when exposed to acid and alkali. KOB was observed to be stable in both acidic and alkaline conditions, although it seems to be

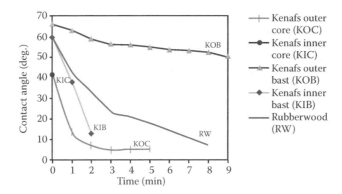

**FIGURE 3.5** Contact angles of 0.1 N HCl solutions on different substrates as a function of time.

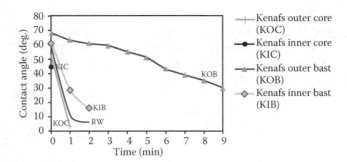

**FIGURE 3.6** Contact angles of 0.1 N NaOH solutions on different substrates as a function of time core.

relatively more sensitive in the latter (as shown by much gradual decrement of contact angle over time). In the case of KIB, KIC, and KOC, they behave like rubberwood irrespective of the level of acidity. Like KOB, the decline in contact angle is much faster in alkaline than in acidic condition. This condition may be attributed to higher amounts of extractives in the bast which acts as a blockage at the openings between the cells, thus penetration of a droplet is inevitably prevented through the surface of the fiber. Abdul Khalil et al. (2010) also reported that the amount of extractive is higher in kenaf bast (5.5%) than in the core (4.7%).

One of the contributing factors for the lack of wetting in KOB fibers is their relatively thicker cell walls. As shown in Figures 3.7a–3.7d, vast differences were observed for the cell wall thicknesses of KOB, KIB, KOC, and KIC parts, respectively. Other studies have reported similar observations; both KOB and KIB showed a wide variation of small-sized lumens (4.60–33.63 μm [Alireza et al. 2003] and 2.80 μm [Abdul Khalil et al. 2010]), while KOC and KIC have bigger sized lumens (8.38–45.65 μm [Alireza et al. 2003] and 6.70 μm [Abdul Khalil et al. 2010]). The presence of vascular bundle gives some advantages for water transportation, or in this case, resins penetration. Conversely, bast fibers may have some difficulties in resin penetration due to the less porous cells available that limit the penetration of the resin adhesive.

### 3.4.7 Buffering Capacity of Kenaf

Table 3.6 represents the pH and buffering capacity values for the extracts of kenaf whole stem (KWS), kenaf core (KC), kenaf bast (KB), and rubberwood (RW). Both pH and buffering capacity are important factors influencing the curing behavior of a resin. The extracts of almost all these materials were found to be acidic. According to Mantanis et al. (2000), all the wood surfaces have a very strong acidic characteristic because the strongest interactions occurred with formamide, which is a basic probe, while much weaker interactions occurred with ethylene glycol, which is an acidic probe. This behavior can be explained by the presence of acidic-type functional groups (specifically, carboxylic acid groups) on the surfaces of the fibers. Studies by many researchers found that most field crops and biomasses are less acidic than wood; the pH of wheat straw was 6.83 (Zhang et al. 2003).

# Adhesion Characteristics of Kenaf Fibers

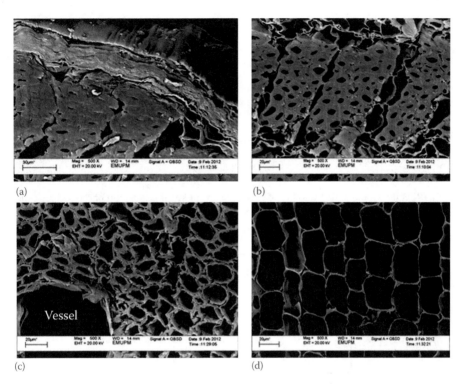

**FIGURE 3.7** SEM of cross sections of kenaf parts under 500× magnification: (a) outer bast, (b) inner bast, (c) outer core, and (d) inner core. (From Juliana, A.H. et al., *J Mater Des* 34:406–411, 2012.)

---

### TABLE 3.6
### The pH and Buffering Capacity Values of Kenaf Whole Stem, Kenaf Core, and Kenaf Bast in Comparison with Rubberwood

| Materials | pH | Acid* 0.1N (HCl) (ml) | Alkali** 0.1N (NaOH) (ml) |
|---|---|---|---|
| Kenaf whole stem (KWS) | 6.55[c] | 15.5[b] | 11.7[a] |
| Kenaf core (KC) | 6.82[b] | 7.4[d] | 2.4[d] |
| Kenaf bast (KB) | 7.94[a] | 24.8[a] | 5.9[c] |
| Rubberwood (RW) | 5.49[d] | 10.5[c] | 6.4[b] |

*Source:* Juliana, A.H. et al., *J Adhes Sci Technol*, 2013.
*Note:* Values are average of samples.
\* The amount of acid needed to reach pH3.0.
\*\* The amount of alkali needed to reach pH10.0 Means followed by the same letters [a,b,c,d] in the same column are not significantly different at $p \leq 0.05$ according to Least Significant Difference (LSD) method.

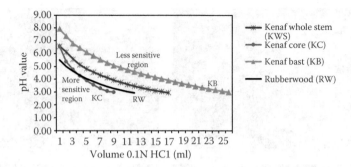

**FIGURE 3.8** Comparative stability of different parts of kenaf stem and rubberwood in acidic liquid. (From Juliana, A.H. et al., *J Adhes Sci Technol*, 2013.)

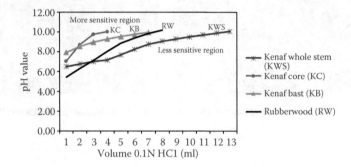

**FIGURE 3.9** Comparative stability of kenaf stem and rubberwood in alkaline liquid. (From Juliana, A.H. et al., *J Adhes Sci Technol*, 2013.)

Figures 3.8 and 3.9 compare the buffering capacities of KWS, KC, KB, and RW in both acidic and alkaline conditions. The choice of pH condition depends on the type of adhesive/matrices use. The study indicates that KWS, KC, and KB are more stable in acidic than in alkaline conditions, since it required more amount of acid to reach pH 3.0 compared to the amount of NaOH required to reach pH 10.0. Whole stem kenaf was observed to be more resistant in both acid and alkali than in separated forms, such as bast and core. KB appears to be more stable in acid as shown by the significantly longer trend lines and a gradual slope (Figure 3.8) which is opposite to that of alkali (Figure 3.9).

Apparently, KWS and RW produced almost a similar trend in acidic conditions. High acidity properties in KWS helps to better cure the aminoplast resins (urea formaldehyde or melamine urea formaldehyde), resulting in relatively higher bond strength. According to Juliana et al. (2012) and Han et al. (2001), better acidity properties of the treated wood or straw particles help improve the bondability between these particles and acid-setting UF resin, and thus the performance of the resulting particleboards. Meanwhile, KC has been observed to be the most sensitive kenaf parts towards both acid and alkali. As seen in both figures, KC has much steeper and shorter slopes thus its buffering capacity is low. The results suggest that bonding of KC with either aminoplast or phenolic resins requires an addition of buffering agent to control the curing rate.

### 3.4.8 BONDING LIMITATION IN KENAF FIBERS

Inorganic adhesion resembles synthetic resin adhesion, where the interfacial bonding between two different materials may involve chemical or physical bonding mechanisms as well as mechanical interlocking (Frihart 2013). The formation of hydrogen bridges can be proven by testing wet and dry composite samples. Dry composites usually showed high strength properties, which can be explained by a high number of hydrogen bonds or hydroxyl bridges. On the other hand, the hydrogen bonds of the wet samples are destroyed by the insertion of water molecules between the bridging hydroxyl groups due to the pressure of the swollen cellulosic fibers, and thus, development of frictional forces between the fibers and binders.

Depending on the cell fracture, the processing of kenafinto particles, fiber, or pulp can create many types of surface attributes, as illustrated in Figure 3.10 (Frihart 2013). When the cell walls are cleaved in a longitudinal trans-wall fashion, the lumen will be the large part and main bonding surface in all natural fibers, especially for kenaf core fibers. The composition of lumen walls can vary between being highly cellulosic to not cellulosic at all, especially if the S3 layer is exposed to high lignin or if they are covered by a warty layer. The middle lamella is also found to be rich in lignin. However, for the most part it is difficult to know when the walls are fractured if the cleavage plane runs through any of the three main fractions or between the lignin-hemicellulose boundaries, which may be the weakest link in the wood cellular structure. The most complicated issue of the bonding surface is the typical mechanical way of providing binding surfaces that cause a lot of fragmentation and smearing of

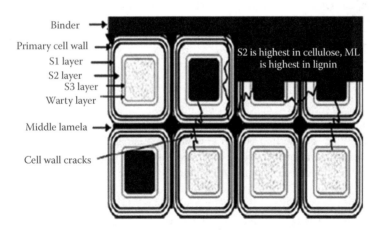

**FIGURE 3.10** Illustration of a transverse section of wood showing fracture points of the wood cellular structure and surfaces available with which binders can interact, assuming clean fractures are occurring. (From Frihart, C.R., Wood adhesion and adhesives, Part III Wood composites, Chapter 9. *Handbook of Wood Chemistry and Wood Composites*. Rowell, R. (Ed.) 2nd Edition, CRC Press, Florida, USA. Pp. 255–320, 2013.)

the cell wall components. The clean splitting of the cell walls can only be viewed by applying a careful microtome sectioning, while other techniques give surfaces that are less intact (Frihart 2013). Thus, the theory of many open lumens into which the adhesive can flow is not always correct, and this can be proven by the fact that penetration of the adhesive into the lumens is not always that fast. The major factors that influence the interaction between fibers and matrix are the matrix composition (cracked or uncracked), the fibers' geometry, the fibers' type, the fibers' surface characteristics, the fibers' orientation, the fibers' volume, and the overall durability of the composite.

### 3.4.9 Surface Inactivation and Adhesive Bonding

Paridah (2013) outlined several important steps that must take place in order for a good bond to occur from the instant an adhesive/matrix is applied to natural fibers (wood and other plant fibers) until the adhesive/matrix cures (i.e., sets into a solid). In the majority of bonding processes, the adhesive/matrix is normally applied onto the surfaces of the fibers using various techniques that suit the manufacturing process and products. Regardless of the technique used, the most important criterion is that the adhesive/matrix must flow and transfer properly to achieve a glue line which uniformly covers the surfaces of the fibers. Also, the adhesive/matrix must penetrate into the fiber surfaces, filling the small voids caused by pores, checks, and other anatomical features. One step further than penetration is wetting. While penetration allows the resin polymer to contact the fiber deep in the crevices on the fiber surface, wetting involves the attraction of the liquid adhesive/matrix to the fiber surface via molecular forces. Finally, the adhesive/matrix must solidify. If any of these actions do not occur, the glue bond will be faulty. Such is the case when gluing fiber which is surface inactivated. The definition, causes, and prevention of surface inactivation in natural fibers (including wood and plant fibers) will be discussed below. Some of the discussion refers to bonding of wood however the explanation can also be applied to other plant fibers.

### 3.4.10 Surface Inactivation

Inactivation of surfaces occurs when the fiber will not allow wetting to happen. An attempt to glue surface-inactivated fiber results in weak glue bonds. The poor bonding can evidently be seen from a broken glue line with either the imprint of the opposite surface and an occasional loose fiber embedded in the glue, such as in the case of laminated fiber composite or protruding fibers or in the case of homogeneous fiber reinforced composites. The appearance of the opposite surface would indicate that penetration took place (the adhesive or matrix filled the voids on the fiber surface), however, if the fiber is surface inactivated, the adhesive or matrix would not be attracted to the fiber molecularly, thus a weak bond occurred at the binder-fiber interface.

To better understand how to prevent problems associated with surface inactivated fibers, a simplified explanation of the wetting phenomenon is presented. The components which make up natural fibers are held together by strong molecular forces. When wood and plant fibers are processed, essentially what happens at the molecular level is the breaking of the bonds between fiber components. Where the molecules once joined are now open bonding sites which possess strong attractive forces (open bonding sites are unstable and molecules desire stability). The higher the number of available bonding sites, the greater will be the total attraction. It is this attraction which gives freshly machined wood and plant fibers its wettability (Frihart 2013). Water, gases, microscopic dust and dirt particles, extractives in the fibers, and adhesives are all likely candidates for open bonding sites. The longer freshly machined wood is exposed to the atmosphere, the more of these bonding sites will be taken by gases and pollutants, and less will be available for the adhesive or matrix. This is why wood and plant fibers lose its wettability over time and the surface becomes inactivated.

Heat is another factor that increases the chance of an inactivated surface. Heat increases the movement of extractives which exist in the natural fibers, increasing the chance that they will move to the fibers' surface and attach to open sites. In addition, severe heat can actually alter the chemistry of fiber components, destroying available bonding sites. Such is the case with wood lumber or veneer that has been dried too harshly, a condition often erroneously referred to as "case hardening" in the wood-based industry.

Fiber may easily be tested for surface inactivation simply by placing a drop of water (using an eye dropper) gently on the surface of the fiber. If the drop sits high on the surface and retains a round shape, surface inactivation is suspected. If the drop sits flat on the surface and begins to take on an oblong shape parallel to the grain, the fiber surface is not inactivated. Another test is to time how long it takes for the drop of water to absorb into the fiber. If the fiber surface is not inactivated, the drop will absorb in several minutes or less. In severely surface inactivated fiber, the water will actually evaporate before it can be absorbed.

## 3.5 THE COMPATIBILITY OF KENAF FIBERS AND POLYMERS

Cellulose, the main component of natural fibers, is completely incompatible with the majority of polymers. This fact has been reported by many studies. Our recent study (Paridah et al. 2012) using an unsaturated-polyester-kenaf-pultruded-composite (Figure 3.11), shows that while the weight of this composite was reduced markedly and the mechanical properties also declined. A significant observation on the failed specimens confirmed the lack of adhesion between the fibers and the matrix as compared to that having 100% glass fibers. Between the polyethylene (PE) and the phenol formaldhyde (PF), the latter appears to have lower wettability. Ku et al. (2011) associated this effect as incompatibility between the hydrophilic natural fibers and the hydrophobic thermoplastic matrices which led to poorer properties of the composites. Malkapuram et al. (2008) asserted that it is necessary to modify the fiber surface by employing chemical modifications to improve the adhesion between fiber and matrix.

## 3.6 FUTURE CHALLENGES

Currently, kenaf fiber polymer composites have two issues, namely resin compatibility and water absorption, that need to be addressed. The research and development

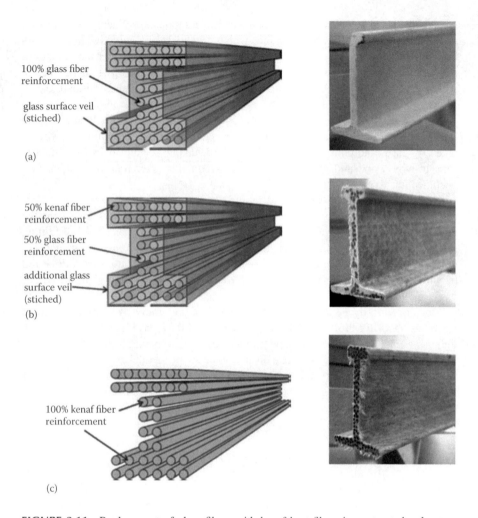

**FIGURE 3.11** Replacement of glass fibers with kenaf bast fibers inunsaturated polyester-pultruded composites: (a) 100% glass fiber, (b) 50% kenaf, and (c) 100% kenaf. (From Paridah Md. Tahir et al., Properties of kenaf reinforced composites in pultrusion and woven lamination. SAMPE Asia 2012. Kuala Lumpur, 2 February, 2012.)

sector had come up with several innovative solutions focusing on these novel materials with specific properties. The use of new additives, e.g., coupling agents or new high-quality preparation methods for natural fibers have enabled more reproducible properties and allowed better processing control. The understanding of the influence of factors such as moisture, fiber type, and fiber content, has improved the composites' mechanical properties and their end-product quality. Moreover, the development of new molding technologies have increased the process efficiency significantly.

# Adhesion Characteristics of Kenaf Fibers

Surface interaction between resin polymer matrices and kenaf fibers is an area that requires more concerted efforts. The characterization of the interface between fiber and matrix provides information on the adhesion strength. This evaluation can be done through several techniques such as:

- Micromechanical techniques: Single fiber pull-out test, fiber bundle pull-out test, single fiber fragmentation test, and micro bond test
- Spectroscopic techniques: Surface characterization of the fiber (before and after treatment)
- Microscopic techniques: SEM, optical microscopy, and stereomicroscopy
- Contact angle measurement: Wettability and surface energy

The second factor limiting large scale production of natural fiber composites is water absorption. Natural fibers absorb water from the air and through direct contact with the environment. This absorption deforms the surface of the composites by swelling and creating voids. The result of these deformations is lower strength and an increase in mass. Additionally, with water absorption rates as high as 20% (w/w) the lightweight advantage is often nullified. Fariborz et al. (2016) studied the relationship between volumetric composition and shear strength of pultruded hybrid kenaf/glass fiber composites and found that the void volume fraction of the composites increases as a function of the kenaf fiber volume fraction at (R2 = 0.95). The study also revealed that the failure in kenaf/fiber glass composites is mainly attributed to the three types of voids in the core region of the composites, namely lumen voids, interface voids, and impregnation voids. The micrographs (Figure 3.12) show that the highest volume of the voids is due to the presence of impregnation voids in the composites. These type of voids were caused by the presence of moisture in kenaf fibers that tried to escape during resin curing, forming a layer of water molecules on the fiber surfaces, and consequently restricting the proper impregnation of the resin polymer into the fibers.

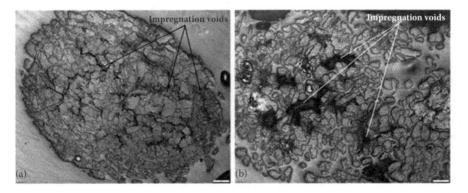

**FIGURE 3.12** Impregnation voids in kenaf/polyester composites with different fiber moisture contents: (a) 0% moisture content of the kenaf fibers; (b) 8.44% moisture content of the kenaf fibers. Scale bar is 100 μm. (From Fariborz, H.D. et al., *J Composite Mater*, 1–13, 2016.)

The treatment of fibers is currently an area of research receiving significant attention. The absorption of water is commonly thought to occur at the free hydroxyl groups on the cellulose chains. With a ratio of 3 hydroxyl groups per glucose repeat unit, the amount of water that can be absorbed is substantial. By capping the hydroxyl groups, this ratio can be reduced. There are several promising techniques that have been studied by various groups (Rowell et al. 2013). Among these treatments, mercerization (alkaline) treatment has had the most reviews. Utilizing silanes as coupling agents is a treatment commonly used in glass composite production and is starting to find uses in natural fiber composites. Acetylation is another treatment that is common with cellulose to form a hydrophobic thermoplastic and has the potential to have the same results on natural fibers.

Today, most of the other natural fiber composites have improved their properties compared to those of several years ago. To satisfy the various needs, many companies offer more additives and auxiliary supplies. Different ones are chosen depending on their applications and compounding methods. In this field, knowledge of chemical interactions (matrix or fiber) and the action mechanisms of additives is necessary. Coupling agents combine fibers and resin, lubricants increase the throughput and improve the surface quality, and colorants and UV stabilizers prevent the polymer degradation and increase visual perception. Biocides can prevent biological decomposition, especially at a higher fiber content. The options are numerous, and it all depends on the industry to choose the method that would give maximum returns to the company.

## REFERENCES

Abdul Khalil H.P.S., A.F. Ireana Yusra, A.H. Bhat, and M. Jawaid. 2010. Cell wall ultrastructure, anatomy, lignin distribution, and chemical composition of Malaysian cultivated kenaf fibre. *Ind. Crops Prod.* 31(1): 113–21.

Ahmed Amel, M. Tahir Paridah, R. Sudin, U.M.K. Anwar, and Ahmed S. Hussein. 2013. Effect of Fibre Extraction Methods on Some Properties of Kenaf Bast Fiber. *Industrial Crops and Products* 46: 117–123.

Aisyah, H.A., M.T. Paridah, M.H. Sahri, A.A. Astimar, and U.M.K. Anwar. 2012. Influence of Thermo-Mechanical Pulping Production Parameters on properties of Medium Density Fibreboard made from Kenaf Bast. *Journal of Applied Science* 12(6): 575–580.

Alireza, A., H. Jalaluddin, and M.Y. Mohd Nor. 2003. Pulping and papermaking properties of Malaysian cultivated kenaf (*Hibiscus cannabinus*). In: *Proceedings of the second technical review meeting on the national project kenaf research project*. Serdang, 108–15.

Anwar, U.M.K., M.T. Paridah, H. Hamdan, A. Zaidon, and E.S. Bakar. 2006. Impregnation of Bamboo (*Gigantochloascortechinii*) Strips with Low-Molecular-Weight Phenol Formaldehyde Resin. *J. Bamboo and Rattan* 15(3–4): 127–133.

Awaja, F., M. Gilbert, G. Kelly, B. Fox, and P.J. Pigram. 2009. Adhesion of polymers. *Progress in Polymer Science* 34: 948–968.

Azizi Mossello, A., Jalaluddin Harun, Hossein Resalati, M. Rushdan Ibrahim, Seyeed Rashid Fallah Shmas, and Paridah Md Tahir. 2010. New approach to use of kenaf for paper and paperboard production. *BioResources* 5(4): 2112–2122.

Biogiotti, J., D. Puglia and J.M. Kenny. 2004. A review on natural fibres based composites—Part 1: Structure processing and properties of vegetable fibres. *Journal of Natural Fibers* 1(2): 37–68.

Bledzki, A.K. and J. Gassan. 1999. Composites Reinforced with Cellulose Based Fibres. *Progress in Polymer Science* 24: 221–274.
Blomquist, R.F., A.W. Christiansen, R.H. Gillespie, and G.E. Myers. 1981. Adhesive Bonding of Wood and Other Structural Materials. Volume III, *Clark C. Heritage Memorial Series on Wood*, Madison, Wis. 12–110.
Calamari, T.A., W. Tao, and W.R. Goynes. 1997. A Preliminary Study of Kenaf Fiber Bundles and Their Composite Cells. *Tappi Journal* 80(8): 149–154.
Chen, L., E.P. Columbus, J.W. Pote, M.J. Fuller and J.G. Black. 1995. *Kenaf bast and core fibre separation*. Kenaf Association, Irving, TX, 15–19.
Chen, M.A., H.Z. Li, and X.M. Zhang. 2007. Improvement of shear strength of aluminium-polypropylene lap joints by grafting maleic anhydride onto polypropylene. *International Journal of Adhesion & Adhesives* 27: 175–187.
Drzal, L.T., A.K. Mohanty, and M. Misra. 2005. *Natural Fibres, Biopolymers and Their Biocomposites*. CRC Press.
Eustathopoulos, N., M.G. Nicholas and B. Drevet (1999). *Wettability at High Temperatures*. Oxford, UK: Pergamon. ISBN 0-08-042146-6.
Fariborz, H.D., M.T. Paridah, B. Madsen, M. Jawaid, D.L. Majid, L. Brancheriau, and A.H. Juliana. 2016.Volumetric composition and shear strength evaluation of pultruded hybrid kenaf/glass fiber composites. *Journal of Composite Materials* 0(0): 1–13. DOI: 10.1177 /0021998315602948.
Freeman, H.G., and F.F. Wangaard. 1960. Effect of wettability of wood on glue-line behavior of two urea resins. *Forest Products Journal*. 10(6): 311–315.
Frihart, C.R. 2013. Wood Adhesion and Adhesives, Part III Wood Composites, Chapter 9. *Handbook of Wood Chemistry and Wood Composites*. Rowell, R. (Ed.) 2nd Edition, CRC Press, Florida, USA, 255–320.
Ghosh, S.K., L.K. Nayak, and S.K. Bhattacharya. 2006. Physicomechanical properties of date palm leaf particle board. *Journal of Agricultural Engineering* 43(4): 89–91.
Juhaida, M.F., M.T. Paridah, M. Mohd. Hilmi, Z. Sarani, H. Jalaluddin, and A.R. MohamadZaki. 2009. Liquefaction of kenaf (*Hibiscus cannabinus* L.) core for wood laminating adhesive. *Bioresorce Technology* 101(1010): 1355–1360.
Juliana A.H., M.T. Paridah, S. Rahim, I. Nor Azowa, and U.M.K. Anwar. 2012. Production of particleboard from kenaf (*Hibiscus cannabinus* L.) as function of particle geometry. *Materials and Design* 34: 406–11.
Juliana, A.H., M.T. Paridah, S. Rahim, I. Nor Azowa, and U.M.K. Anwar. 2013. Affect of adhesion and properties of kenaf (*Hibiscus cannabinus* L.) stem in particleboard performance. *Journal of Adhesion Science and Technology*. http://dx.doi.org/10.1080 /01694243.2013.848622.
Juliana, A.H., M.T. Paridah, M.H. Sahri, U.M.K. Anwar, and A.A. Astimar. 2012. Properties of particleboard made from kenaf (*Hibiscus cannabinus* L.) as function of particle geometry. *Journal Materials and Design* 34: 406–411.
Jusoh, M.J. 2008. Studies on the properties of woven natural fibers reinforced unsaturated polyester composites. *Master of Science thesis*, Universiti Sains Malaysia, Penang, Malaysia.
Ku, H., H. Wang, N. Pattarachaiyakoop, and M. Trada. 2011. A review on the tensile properties of natural fibre reinforced polymer composites. Centre of Excellence in Engineered Fibre Composites and Faculty of Engineering, University of Southern Queensland. eprints.usq.edu.au/18884/1/Ku_Wang_ Pattarachaiyakoop_Trada_AV.pdf
Malkapuram, R., V. Kumar, and S.N. Yuvraj. 2008. Recent Development in Natural Fibre Reinforced Polypropylene Composites, *Journal of Reinforced Plastics and Composites* 28, 1169–1189.
Mantanis, G., P. Nakos, J. Berns, and L. Rigal. 2000. Turning agricultural straw residues into value-added composite products: A new environmentally friendly technology. Proceedings of the 5th International Conference on Environmental Pollution.

Anagnostopoulos, A. (Ed.) Aug. 28–31. University of Thessaloniki. Thessaloniki, Greece, 840–848.

Marra, A.A. 1992. *Technology of Wood Bonding*. Van Nostrand Reinhold, New York, NY.

Moradbak, A., M.T. Paridah, A.Z. Mohamed, and R.B. Halis. (2015). Alkaline Sulfite Anthraquinone and Methanol Pulping of Bamboo (*Gigantochloascortechinii*). *BioResources* 11(1): 235–248.

Naji, H.R., M.H. Sahri, T. Nobuchi, and E.S. Bakar. 2012. Clonal and planting density effects on some properties of rubber wood (*Hevea brasiliensis* Muell. Arg). *BioResources Journal* 7(1): 189–202.

Nayeri, M.D., M.T. Paridah, J. Harun, L.C. Abdullah, E.S. Bakar, M. Jawaid, and F. Namvar. 2013. Effects of Temperature and Time on the Morphology, pH, and Buffering Capacity of Bast and Core Kenaf Fibres. *BioResources* 8(2): 1802–1812.

Norul Izani, M.A., M.T. Paridah, U.M.K. Anwar, M.Y. Mohd Nor, and P.S. H'ng. 2013. Effect of fibre treatment on morphology, tensile and thermogravimetric analysis of oil palm empty bunches fibres. *Composites Part B*, 45, 1251–1257.

Packham, D.E. 2011. Theories of Fundamental Adhesion. Lucas F.M. da Silva, Andreas Ochsner, Robert D. Adams (Eds.), *Handbook of Adhesion Technology*, DOI 10.1007 /978-3-642-01169-6_2, # Springer-Verlag Berlin Heidelberg.

Paridah Md. Tahir. 2013. *Bonding with Natural Fibres*. Inaugural Series. Universiti Putra Malaysia. UPM Press, 112.

Paridah Md. Tahir, Amel B. Ahmed, Syeed, O.A. Saiful, Azry and Zakiah Ahmed. 2011. Retting Process of Some Plant Fibres and Its Effect on Fibre Quality: A Review. *BioResources* 6(4): 5260–5281.

Paridah Md. Tahir, V. Klapper, Edi Syam Zainuddin, and Khalina Abdan. 2012. Properties of kenaf reinforced composites in pultrusion and woven lamination. SAMPE Asia 2012. Kuala Lumpur, 2 February 2012.

Paridah, M.T., A.W. Nor Hafizah, A. Zaidon, I. Azmi, M.Y. Mohd. Nor, and M.Y. Nor Yuziah. 2009. Bonding Properties and Performance of Multilayered Kenaf Board. *Journal of Tropical Forest Science* 21(2): 113–122.

Paridah, M.T., and A. Khalina. 2009. Effects of soda retting on the tensile strength of kenaf (*Hibiscus cannabnius* L.) bast fibres. *Project Report Kenaf EPU*, 21.

Paridah, M.T., S.A. Syeed, and H. Juliana. 2008. Mechanical properties of kenaf (*Hibiscus cannabnius* L.) stem of different ages. Proceedings Colloquium of Kenaf Research Output 1–2 December 2008, Seremban, Negeri Sembilan Malaysia Unversiti Putra Malaysia, 18.

Rowell, R.M. and J.S. Han. 1999. Changes in kenaf properties and chemistry as a function of growing time. *Kenaf Properties, Processing and Products*. Mississippi State University. Mississippi State, MS., 32–57.

Rowell, R.M., R. Pettersen, and M.A. Tshabalala. 2013. Cell wall Chemistry. Chapter 3, *Handbook of Wood Chemistry and Wood Composites*. R.M. Rowell. (Ed). 2nd Edition, CRC press, Taylor & Francis Group, Boca Raton, FL.

Rowell, R.M. and H.P. Stout. 2006. Jute and Kenaf. *Handbook of Fibre Chemistry*, Menachem Lewin (Ed.). CRC Press.

Sharfrin, E. and W.A. Zisman. (1960). Constitutive relations in the wetting of low energy surfaces and the theory of the retraction method of preparing monolayers. *Journal of Physical Chemistry* 64(5): 519–524. DOI:10.1021/j100834a002.

Wahab, N.H.A., P.M. Tahir, Y.B. Hoong, Z. Ashaari, N.Y.M. Yunus, M.K.A. Uyup, and M.H. Shahri. 2012. Adhesion characteristics of phenol formaldehyde pre-preg oil palm stem veneers. *BioResources* 7(4): 4545–4562.

Zhang, T. 2003. Improvement of kenaf yarn for apparel application. *Master's thesis*, Louisiana State University, US.

# 4 Kenaf Fiber-Reinforced Thermoplastic Composites

*A. Khalina and N. Mohd Nurazzi*

## CONTENTS

4.1 Introduction ........................................................................................................ 61
    4.1.1 Kenaf Fiber ............................................................................................. 62
    4.1.2 Kenaf Fiber-Reinforced Thermoplastic Polymers ................................. 63
4.2 Critical Role Effecting Performance of Natural Fiber and Kenaf
    Fiber-Reinforced Polymer Composites ........................................................... 69
4.3 Processing Method for Kenaf and Other Natural Fiber-Reinforced
    Thermoplastic Composites ............................................................................... 71
    4.3.1 Compression Molding ........................................................................... 71
    4.3.2 Extrusion ................................................................................................ 72
    4.3.3 Injection Molding .................................................................................. 72
4.4 Characterization of Kenaf-Reinforced Thermoplastic Composites ................ 72
    4.4.1 Kenaf-Reinforced Polypropylene Composites ..................................... 72
    4.4.2 Kenaf-Reinforced Low Density Polypropylene Composites .............. 73
    4.4.3 Kenaf-Reinforced Polyurethane Composites ....................................... 73
    4.4.4 Kenaf-Reinforced Polylactic Acid (PLA) Composites ....................... 75
References ................................................................................................................. 76

## 4.1 INTRODUCTION

Recently, the use of thermoplastic composites produced from synthetic polymers reinforced with natural fibers has attracted the attention of many researchers and industries worldwide. Generally, the promising performance of synthetic fibers, such as glass and carbon fibers, in the manufacturing of composites, possess a higher specific strength and modulus; hence, enabling them as valuables material in a huge number of industrialized requests which requires those features (Nunna et al. 2012).

In spite of all those excellent mechanical properties, these materials may cause environmental pollution due to the non-degradability of fibers, and are unable to be incinerated as their residues tend to cause furnace damage and other problems associated with the recycling of glass fiber-reinforced thermoplastics due to fiber breakages occurring during reprocessing operations. The only common method of their disposal is to discard the waste in landfills, which is becoming more costly in many countries with the introduction of landfill, machine, labor, and operation costs (Bos, 2004).

As an alternative solution to the ever-depleting petroleum sources, natural fibers can be used as a material for composite-based products, and are receiving greater attention from researchers, industries and communities. Natural fiber-based composites are attractive due to their biodegradability, signifying strength with its density ratio, nontoxicity, and relative strength, making them ideal products which can be used in the construction, automotive, and furniture industries (Paukszta and Borysiak, 2013; Sapuan et al., 2013).

An innumerable demerit of natural fiber is hydrophilic in nature, with poor fiber to matrix interfacial adhesion and poor thermal stability. This issue can be overcome by a modification treatment and compatibilizer which will amend the surface and structure which hence enhance the interfacial adhesion between the fiber and matrix. Natural fiber-reinforced thermoplastic composites possess the variances and incomparability in terms of their polarity structures (Rassiah and Megat Ahmad, 2013). Other than that, the consideration of natural fibers is dependant on the crystalline content, size, shape, orientation, and thickness of their cell walls, and effect of the fiber volume fraction. These different structures are reflected in the stress–strain diagrams of the fibers (Shalwan and Yousif, 2013).

Kenaf fiber is one of the celebrated and gratifying increasingly constituents of natural fiber-reinforced thermoplastic composites throughout the world and even in Malaysia. The research on kenaf-reinforced thermoplastic received much attention owing to its significance properties for a wide variety of applications, such as the development of eco-friendly assets for the automotive parts, sports, food packaging, furniture (Anuar and Zuraida, 2011), textiles, paper and pulp, and fiberboard industries (Saba et al., 2015). Inferior thermal resistance is displayed by kenaf when compared to synthetic fibers, such as carbon and glass fibers, like all other natural fibers (Azwa and Yousif, 2013).

Due to the good mechanical strength and thermal properties of kenaf composites, thus it is suitable suitable applicant for high performance material such as extruded plastic fencing, decking, and furniture padding (Thiruchitrambalam et al., 2009).

### 4.1.1 KENAF FIBER

Kenaf fiber (*Hibiscus cannabinus*) is obtained from the stems of plants. It is a herbaceous annual plant type with a short growing period, where it can yield three times a year in Malaysian climates (Norlin et al. 2011). The stem of kenaf tree consists of a bast and core, and it produces approximately 34 to 38% of bast fibers (Aji et al., 2009). According to Paridah et al. (2011), the bast of kenaf fibers possesses good mechanical and physical properties and can replace glass fiber composites as reinforcement.

Many researchers have reported that kenaf fiber as reinforcement a showed significantly improved properties in polymer composites beneath tensile stress, flexural, and impact loading circumstances when compared to other natural fibers as reinforcement. It reported that the major constituents of natural fibers are cellulose and lignin

# Kenaf Fiber-Reinforced Thermoplastic Composites

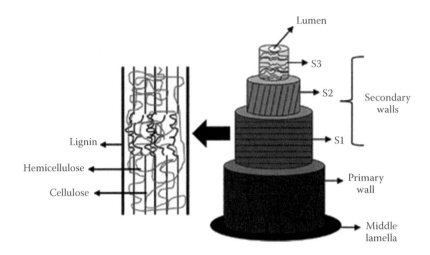

**FIGURE 4.1** Schematic position of main constituent of natural fibers. (From Pereira, P.H.F. et al., *Polímeros*, 25, 9–22. 2015.)

(see Figure 4.1). Cellulose as a main structural component plays an important role to provide mechanical strength to the fibers. Thus, for lignin it would give rigidity to the plant, affecting the modulus of the fibers. According to Periera et al. (2015), there is about 72.0% of cellulose, 20.3% of hemicellulose, 9.0% of lignin, and 4.0% of ash contained in the structure of kenaf fibers (Pereira et al., 2015).

Table 4.1 shows that comparison of mechanical properties of kenaf fiber with other natural and synthetic fibers that widely used in polymer composites (Ishak et al., 2013; Kestur and Fernando, 2009; Koronis and Fontul, 2013; Ku et al., 2011).

### 4.1.2 Kenaf Fiber-Reinforced Thermoplastic Polymers

One of the big new areas of development is in combining natural fibers with thermoplastics. Since the process for thermoplastic materials have risen sharply over the past few years, adding natural fiber to thermoplastic provides a cost reduction and give some significant increases in performance. According to Rowell et al. (1999), there are two concentrated research areas of natural fiber with thermoplastic polymers; one in which no attempt is made to compatibilize the two dissimilar resources, and a second in which the compatibilizer is used to make the hydrophobic nature of thermoplastic cooperate well with the hydrophilic nature of natural fibers.

Besides consideration on the compatibility of natural fiber, especially kenaf and thermoplastic, natural fibers have a lower processing temperature due to the possibility of fiber degradation and volatile emission that could affect composite performance. The processing temperature is thus limited to about 200°C, although it is possible to use higher temperature for short periods. This processing factor limits to a certain type commodity thermoplastic such as polyethylene (PE), polypropylene (PP), and polystyrene (PS) (Rowell et al., 1999).

## TABLE 4.1
### Mechanical Properties of Selected Natural Fibers and Synthetic Fibers

| Fibers | Density (g/cm$^3$) | Tensile Strength (MPa) | Elongation at Break (%) | Young Modulus (GPa) |
|---|---|---|---|---|
| Sugar Palm | 1.29 | 15.5–290 | 5.7–28.0 | 0.5–3.37 |
| Bagasse | 1.5 | 290 | – | 17 |
| Bamboo | 1.25 | 140–230 | – | 11–17 |
| Flax | 0.6–1.1 | 345–1035 | 2.7–3.2 | 27.6 |
| Hemp | 1.48 | 690 | 1.6–4 | 70 |
| Jute | 1.3 | 393–773 | 1.5–1.8 | 26.5 |
| Kenaf | 1.45 | 930 | 1.6 | 53 |
| Sisal | 1.5 | 511–535 | 2.0–2.5 | 9.4–22 |
| Ramie | 1.5 | 560 | 2.5–3.8 | 24.5 |
| Pineapple | 0.8–1.6 | 400–627 | 14.5 | 1.44 |
| Coir | 1.2 | 175 | 30 | 4–6 |
| E-Glass | 2.5 | 2000–3500 | 0.5 | 70 |
| S-Glass | 2.5 | 4570 | 2.8 | 86 |
| Aramid | 1.4 | 3000–3150 | 3.3–3.7 | 63.0–67.0 |

*Sources:* Ishak, M.R. et al., *Carbohydrate Polymers, 91*(2), 699–710, 2013; Kestur, G.S.G. et al., *Progress in Polymer Science, 34*(9), 982–1021, 2009; Koronis, G.S., A. Fontul, M., *Composites Part B: Engineering, 44*(1), 120–127, 2013; Ku, H.W. et al., *Composites Part B: Engineering, 42*(4), 856–873, 2011.

There are five important roles of matrix; to hold the reinforcement phase in place, to deform and distribute the stress to the reinforcement under applied loads or stress, to bind the fibers together and transfer the load to the fibers and provide rigidity and shape to the structure, to isolate the fibers so that individual fibers can act separately and stops or slow the propagation of cracks, and to provide protection to the reinforcing fibers against chemical attack and mechanical damage (Callister, 2012; Mazumdar, 2001).

According to Holbery and Houston (2006), in automotive applications, the most common system used today is thermoplastic polypropylene, especially in nonstructural components. Figure 4.2 shows the inner door fabricated from kenaf-reinforced polypropylene composites. Polypropylene is favored due to its low density, excellent processing ability, good mechanical properties, and dimensional stability. Other than polypropylene, some commonly used composites are polyethylene, polystyrene, polyurethane and polyamide (nylon 6 and 6, 6). Table 4.2 presents different type of thermoplastic polymers mostly employed in fabricating natural fiber polymer composites for various applications (Holbery and Houston, 2006).

Thermoplastic materials are also generally good insulating materials (385 to 1775 V/mil). The effective of insulating strength of thermoplastic is measured by the

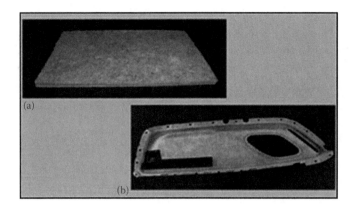

**FIGURE. 4.2** Inner door from kenaf-reinforced polypropylene composites. (a) Kenaf mat (b) Inner door (50% kenaf, 50% polypropylene.) (From Mohanty, A.K.M., M. Drzal, L.T., *Natural Fibers, Biopolymers, and Biocomposites*: Taylor & Francis, 2005.)

### TABLE 4.2
### Typical Thermoplastic Polymers Used in Natural Fiber Composite Fabrication

| Property | PP | LDPE | HDPE | PS |
|---|---|---|---|---|
| Density (g/cm$^3$) | 0.899–0.920 | 0.910–0.925 | 0.94–0.96 | 1.04–1.06 |
| Water absorption (24 h; %) | 0.01–0.02 | <0.015 | 0.01–0.2 | 0.03–0.10 |
| Tg (°C) | −10 to −23 | −125 | −133 to −100 | – |
| Tm (°C) | 160–176 | 105–116 | 120–140 | 110–135 |
| Heat deflection temp (°C) | 50–63 | 32–50 | 43–60 | Max. 220 |
| Tensile strength (MPa) | 26–41.4 | 40–78 | 14.5–38 | 25–69 |
| Tensile modulus (GPa) | 0.95–1.77 | 0.055–0.38 | 0.4–1.5 | 4–5 |
| Elongation (%) | 15–700 | 90–800 | 20–130 | 1–2.5 |
| Izod impact strength (J/m) | 21.4–267 | >854 | 26.7–1068 | 1.1 |

*Source:* Holbery, J., Houston, D., *Jom*, 58(11), 80–86, 2006.
*Note:* Polypropylene (PP), low density polypropylene (LDPE), high density polypropylene (HDPE), polystyrene (PS).

dielectric strength, where it defined as the voltage gradient that produces electrical breakdown through the material (Smith and Hashemi, 2006).

Table 4.3 showed the list of recent works of researchers on kenaf reinforced thermoplastic composite. According to Rassiah et al. (2013), the mechanical properties of kenaf reinforced composites may vary due to different testing methods, the sample tested (Rassiah and Megat Ahmad, 2013), sources of fibers and grade of resin used.

## TABLE 4.3
## Several Recent Works Had Been Done by Researcher on Kenaf Fiber-Reinforced Thermoplastic Polymer Composites in the Past 2009–2015

| No | Research Area | Matric | Year | References |
|---|---|---|---|---|
| 1 | Effect of Kenaf Fibre modification on morphology and mechanical properties of thermoplastic polyurethane materials | TPU | 2015 | (Datta and Kopczyńska, 2015) |
| 2 | The effect of binder on mechanical properties of Kenaf Fibre/polypropylene composites using full factorial method | PP | 2014 | (Amran et al., 2015) |
| 3 | Fiber dispersion during compounding/injection molding of PP/Kenaf Composites: Flammability and mechanical properties | PP and MAPP | 2015 | (Subasinghe et al., 2015) |
| 4 | Modification of Nano-Kenaf surface with maleic anhydride grafted polypropylene upon improved mechanical properties of polypropylene composite | PP | 2015 | (Kim, 2015) |
| 5 | Reinforcing mechanical, water absorption and barrier properties of poly(Lactic acid) composites with Kenaf-derived cellulose of thermally-grafted aminosilane | PLA | 2015 | (Tee et al., 2015) |
| 6 | Effects of Polyethylene-G-Maleic anhydride on properties of low density polyethylene/thermoplastic sago starch reinforced Kenaf Fibre composites | LDPE | 2011 | (Rohani, 2011) |
| 7 | Adhesive wear of thermoplastic composite based on Kenaf Fibres | TPU | 2011 | (Narish et al., 2011) |
| 8 | Effect of fiber treatment on mechanical properties of Kenaf Fiber-Ecoflex composites | Ecoflex (biodegradable thermoplastic) | 2009 | (Nor Azowa et al., 2009) |
| 9 | The effect of processing parameters on the mechanical properties of Kenaf Fibre plastic composite | PP | 2011 | (Bernard et al., 2011) |
| 10 | Influence of fiber content on the mechanical and thermal properties of Kenaf Fiber reinforced thermoplastic polyurethane composites | TPU | 2012 | (El-Shekeil et al., 2012) |

(*Continued*)

## TABLE 4.3 (CONTINUED)
## Several Recent Works Had Been Done by Researcher on Kenaf Fiber-Reinforced Thermoplastic Polymer Composites in the Past 2009–2015

| No | Research Area | Matric | Year | References |
|---|---|---|---|---|
| 11 | Effect of alkali treatment on mechanical and thermal properties of Kenaf Fiber-Reinforced thermoplastic polyurethane composite | TPU | 2012 | (El-Shekeil et al., 2012) |
| 12 | Development of green insulation boards from Kenaf Fibres and polyurethane | TPU | 2011 | (Ibraheem et al., 2011) |
| 13 | Poly(Lactic Acid) (PLA)-reinforced Kenaf bast fibers composites: The effect of triacetin | PLA | 2009 | (Ibrahim et al., 2009) |
| 14 | Effects of extrusion temperature on the rheological, dynamic mechanical and tensile properties of Kenaf Fiber/HDPE composites | HDPE | 2014 | (Salleh et al., 2014) |
| 15 | Renewable resource based "All Green Composites" from Kenaf Biofiber and Poly(Furfuryl alcohol) bioresin | Poly(furfuryl alcohol) bioresin | 2013 | (Deka et al., 2013) |
| 16 | The effects of alkali–silane treatment on the tensile and flexural properties of short Fibre non-woven Kenaf reinforced polypropylene composites | PP | 2012 | (Asumani et al., 2012) |
| 17 | Environmental performance of Kenaf-Fiber Reinforced polyurethane: A life cycle assessment approach | Rigid PU | 2014 | (Batouli et al., 2014) |
| 18 | Tensile properties of Kenaf Fiber and corn husk flour reinforced poly(Lactic acid) hybrid bio-composites: Role of aspect ratio of natural fibers | PLA | 2014 | (Kwon et al., 2014) |
| 19 | Kenaf/Polypropylene nonwoven composites: The influence of manufacturing conditions on mechanical, thermal, and acoustical performance | PP | 2013 | (Hao et al., 2013) |
| 20 | Effect of ammonium polyphosphate on flame retardancy, thermal stability and mechanical properties of alkali treated Kenaf Fiber filled PLA biocomposites | PLA | 2014 | (Shukor et al., 2014) |

*(Continued)*

## TABLE 4.3 (CONTINUED)
## Several Recent Works Had Been Done by Researcher on Kenaf Fiber-Reinforced Thermoplastic Polymer Composites in the Past 2009–2015

| No | Research Area | Matric | Year | References |
|---|---|---|---|---|
| 21 | Flammability, biodegradability and mechanical properties of bio-composites waste polypropylene/Kenaf Fiber containing nano Caco$_3$ with diammonium phosphate | HDPE | 2012 | (Suharty et al., 2012) |
| 22 | Kenaf–Polypropylene composites: Effect of amphiphilic coupling agent on surface properties of fibres and composites | PP | 2012 | (John et al., 2010) |
| 23 | Mechanical properties of Kenaf/polypropylene nonwoven composites | PP | 2010 | (Hao et al., 2012) |
| 24 | Mechanical properties of soil buried Kenaf Fibre reinforced thermoplastic polyurethane composites | TPU | 2013 | (Sapuan et al., 2013) |
| 25 | Studying the effect of fiber size and fiber loading on the mechanical properties of hybridized Kenaf/PALF-Reinforced HDPE composite | HDPE | 2011 | (Aji et al., 2011) |
| 26 | Effect of filler loading and naoh addition on mechanical properties of moulded Kenaf/Polypropylene composite | PP | 2015 | (Md Radzi et al., 2015) |
| 27 | Effects of Kenaf loading on processability and properties of linear low-density polyethylene/poly (vinyl alcohol)/Kenaf composites | LLDPE | 2015 | (Ai Ling et al., 2015) |
| 28 | Influence of chemical treatment on the tensile properties of Kenaf Fiber reinforced thermoplastic polyurethane composite | TPU | 2012 | (El-Shekeil et al., 2012) |
| 29 | Development of a new Kenaf Bast Fiber-Reinforced thermoplastic polyurethane composite | TPU | 2011 | |
| 30 | Optimization of compression moulding temperature for polypropylene materials | PP | 2011 | (Maringgal et al., 2011) |
| 31 | Kenaf performance in PP/EVA/Clay biocomposite | PP | 2011 | (Kamaruddin et al., 2011) |

*Note:* Polypropylene (PP), low density polyethylene (LDPE), linear low density polyethylene (LLDPE), high density polyethylene (HDPE), thermoplastic polyurethane (TPU), polylactic acid (PLA), maleic anhydride-grafted *polypropylene (MAPP)*, ethylene Vinyl Acetate (EVA).

## 4.2 CRITICAL ROLE EFFECTING PERFORMANCE OF NATURAL FIBER AND KENAF FIBER-REINFORCED POLYMER COMPOSITES

There are several factors contributing by natural fibers which influence to the performance of composites such as fiber loading, fiber length, fiber orientation, chemical structure, and interfacial adhesion between fiber and matrix (Fu et al., 2008; Ochi, 2008; Rassiah and Megat Ahmad, 2013; Saba et al., 2015; Shalwan and Yousif, 2013) According to Shalwan et al. (2013), among all the factors, the mechanical efficiency of natural fiber-reinforced polymer composites depend on the fiber-matrix interfacial adhesion where it relates to the ability to transfer stress from the matrix to the fiber (Chin and Yousif, 2009; Mohd Nurazzi et al., 2017; Shalwan and Yousif, 2013).

The interfacial adhesion between fiber and matrix are highly dependent on each other because of the natural fibers; especially kenaf, which are rich in cellulose, hemicellulose, pectins, and lignin, which are hydroxyl groups. In other words, there will be a significant problem in compatibility which weakens the interface area between natural fibers (strong polar and hydrophilic) and polymer matric (non-polar and hydrophobic). The hydrophilic nature of natural fibers leads the composites having high water absorption characteristics that reduce their performance to the composites. Table 4.4 showed the main composition of certain natural fibers (Li et al., 2007).

### TABLE 4.4
### Chemical Composition of Some Common Natural Fibers

| Fiber | Cellulose (wt%) | Hemicellulose (wt%) | Lignin (wt%) | Waxes (wt%) |
|---|---|---|---|---|
| Bagasse | 55.2 | 16.8 | 25.3 | – |
| Bamboo | 26–43 | 30 | 21–31 | – |
| Flax | 71 | 18.6–20.6 | 2.2 | 1.5 |
| Kenaf | 72 | 20.3 | 9 | – |
| Jute | 61–71 | 14–20 | 12–13 | 0.5 |
| Hemp | 68 | 15 | 10 | 0.8 |
| Ramie | 68.6–76.2 | 13–16 | 0.6–0.7 | 0.3 |
| Abaca | 56–63 | 20–25 | 7–9 | 3 |
| Sisal | 65 | 12 | 9.9 | 2 |
| Coir | 32–43 | 0.15–0.25 | 40–45 | – |
| Oil palm | 65 | – | 29 | – |
| Pineapple | 81 | – | 12.7 | – |
| Curaua | 73.6 | 9.9 | 7.5 | – |
| Wheat straw | 38–45 | 15–31 | 12–20 | – |
| Rice husk | 35–45 | 19–25 | 20 | 14–17 |
| Rice straw | 41–57 | 33 | 8–19 | 8–38 |

*Source:* Li, X. et al., *Journal of Polymers and the Environment*, 15(1), 25–33, 2007.

The natural presence of hydroxyl groups in the amorphous region in the lignin substance will reduce the adhesion of natural fibers with most of the matrices, and thus lead to dimensional instability and poor mechanical properties. Furthermore, the high cellulose content of natural fibers also contains a high crystalline content. The crystalline regions are an aggregation of cellulose block which are held together closely by the strong intramolecular hydrogen bonds.

The high degree of crystallinity of cellulose and the tri-dimensional reticulate's structure of lignin make natural fibers far from their thermoplastic counterparts, thus the crystalline structure of cellulose can be disrupted by substituting the hydroxyl group (-OH) with some chemical functionality. Besides that, the presence of a natural waxy layer on the fiber surface may hinder the matrix's penetration into the fiber structure (Mohanty et al., 2001).

Therefore, in order to improve and develop the natural fiber-reinforced polymer composites with better mechanical, physical, and thermal properties of the composites, it is necessary to improve the hydrophobicity nature of natural fibers, impurities, and interfacial adhesion of the fiber and matrix by modification, relying on physical and chemical treatments (La Mantia and Morreale, 2011).

The physical treatments include corona treatment and plasma treatment, while the chemical treatments include alkali treatment (mercerization), silane treatment, acetylation, etherification, acrylation, peroxide treatment, sodium chloride (HCl) treatment, titanate treatment, permanganate treatment, etc. From all the reported treatments and modifications done by the researchers, the alkali treatment (mercerization) was the most prominent one due to its easily implementation, cost-effectiveness, abundance of chemical sources (e.g., NaOH) and contribution to significant improvement on mechanical properties (Bachtiar and Hamden, 2008; Faruk et al., 2012; Leman et al., 2008; Sreekala et al., 2000).

The alkali treatment involved the process of cleaning impurities from the surface of the fibers, which, in turn, increased the roughness of the fiber surface and disrupted the moisture absorption process through removing the coat of -OH groups in the fiber, as illustrated in Figure 4.3. Several works report that NaOH is able to enhance the surface morphology of fibers and increase the number of pores in the fiber surface with the increase of concentration. Furthermore, the use of NaOH in all conditions had resulted a rougher surface and better separation in removing impurities, non-cellulosic materials, inorganic substances, and wax. Then the range representing of -OH stretching of hydrogen bonds became less intense upon the alkali treatment, and hence improved interface linkage or fiber-matrix adhesion compared to the untreated fiber. Figure 4.4 shows the mechanism on how the coupling agent may contribute to the improvement of the interfacial adhesion between the natural fibers with polymer matrix.

$$\text{Natural fiber-OH} + \text{NaOH} \longrightarrow \text{Natural fiber-O}^- \text{Na}^+ + \text{N}_2\text{O}$$

**FIGURE 4.3** Schematic reaction of natural fibers with NaOH solution (alkali treatment).

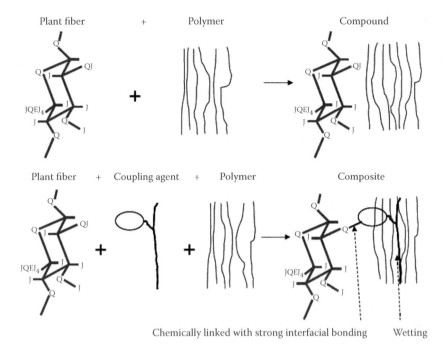

**FIGURE 4.4** Mechanism of coupling agent between hydrophilic fiber and hydrophobic matrix polymer. (From Liu, W.M. et al., *Polymer*, 45(22), 7589–7596, 2004.)

## 4.3 PROCESSING METHOD FOR KENAF AND OTHER NATURAL FIBER-REINFORCED THERMOPLASTIC COMPOSITES

### 4.3.1 COMPRESSION MOLDING

Compression molding was among the earliest method of molding to be used to produce plastic components. It is a pressure method that very popular in the manufacturing of natural fiber composites because of its high reproducibility and low cycle time. In the compression molding process, flat, semi-finished products or hybrid fleeces are usually used, which are either larger than the form or are cut exactly to the size of the desired dimensions. The material placed in between the molding plates flows due to the application of pressure and heat and acquires the shape of the mold cavity and mold design. Pressure, temperature, and time are critical parameters and have to be optimized effectively in order to achieve the desired dimensions and specifications. Separate platens can be used to solve this problem, where one is hot press that is electrically heated, and another one is cold press that is water cooled. The hot pressed component and then transferred immediately to the cold press to chill it under pressure. The hot press is usually preheated to reduce the total cycle time.

## 4.3.2 Extrusion

The extrusion process is used by the plastic industry for mixing the additives with the thermoplastic materials and finally for the production of granules in a larger scale. Single or twin screws that run either by corotating or counter-rotating may be used for this process. Single screw extruders are used when the mixing effect does not have to be very high. While for the twin screw extruder, the excellent mixing effect of the compound can be homogenously distributed due to intermeshing corotating twin screw extruders hence leads to homogeneity output in the close barrel.

## 4.3.3 Injection Molding

Injection molding is by far the most used processing technique of producing parts from thermoplastic materials due to its high productivity. Injection molding machines and molds are very expensive because of the high pressures required and complexity of the process control. However, the shortcomings of this technique are balanced by its ability to produce a complex finished part in a single and rapid operation. The principle of injection molding is very simple. The plastic material is fed into the injection barrel by gravity through the hopper. Upon entrance into the barrel, the polymer is heated to the melting temperature. It is then forced into a closed mold that defined by the shape of the article to be produced. The mold is cooled constantly to a temperature that allows the molten to solidify; when the mold is opened, the finished product is ejected and the process continues.

## 4.4 CHARACTERIZATION OF KENAF-REINFORCED THERMOPLASTIC COMPOSITES

### 4.4.1 Kenaf-Reinforced Polypropylene Composites

Asumani et al. (2012) have studied the effect of alkaline treatments on tensile and flexural properties of kenaf-reinforced PP composites. The mechanical and SEM analysis showed kenaf fiber treated with NaOH solution and NaOH with three-aminopropyltriethoxysilane were significantly increased the tensile and flexural properties which reinforced PP composites. 5% concentration of NaOH with 30% of kenaf fiber loading is the optimum tensile strength and flexural strength, while for NaOH concentration of 6% is the optimum for NaOH with three-aminopropyltriethoxysilane treatment (Asumani et al., 2012).

Furthermore, Bernard et al. (2011) investigates the effects that processing parameters, including temperature (190°C, 200°C, 210°C, 220°C, 230°C, and 240°C) and speed (12Hz, 16Hz, 20Hz, and 24Hz) on the mechanical and thermal properties of unidirectional kenaf fiber-reinforced PP composite. The optimal processing parameters using developed compression molding machine are a temperature of 230°C and a barrel speed of 16Hz. The high tensile strength of the PP/kenaf composite was 35.1MPa. In contrast, the impact properties reduced to 130.9J/M for the unnotched izod test and 79.5J/M for the notched izod test. This result is obviously different from that with pure PP, where the value is 328J/M. Dynamic mechanical analysis (DMA) indicates that the storage modulus and loss modulus for UKF reinforced PP has better properties than pure PP (Bernard et al., 2011).

Hao et al. (2013) have studied the influence of manufacturing conditions on mechanical, thermal, and acoustical (sound absorption and sound transmission loss) performance of kenaf/polypropylene nonwoven composites (KPNCs). Researcher found that temperature and time are the most significant processing factors. For the 6-mm KPNC samples, processing at 230°C for 120s (sample 5/230/120 or 7/230/120) gave the best mechanical properties. In contrast, samples 5/200/60 and 7/200/60, was the lowest modulus were the best impact energy and sound absorbers and excellent sound barriers due to their panel-felt-panel sandwich structure. In addition, the manufacturing conditions did not significantly affect the thermal properties of composites. KPNCs were more thermally stable than virgin PP plastics by adding kenaf fiber as reinforcement. The sound insulation test verified that the classic panel resonance theory could be applied to the KPNCs produced in this study (Hao et al., 2013).

### 4.4.2 KENAF-REINFORCED LOW DENSITY POLYPROPYLENE COMPOSITES

The tensile strength and Young's modulus of LDPE/thermoplastic sago starch (TPSS)–kenaf fiber composites with the addition of polyethylene-g-maleic anhydride (PE-g-MA), showed greater strength than the composites without the addition of PE-g-MA particularly at higher fiber loading. The interfacial properties between LDPE, TPSS, and kenaf fibers were improved after the addition of PE-g-MA. Equilibrium water uptakes for LDPE/TPSS–kenaf fiber composites were higher than those of LDPE/TPSS–kenaf fiber composites with the addition of PE-g-MA, due to lower abundance of hydrogen group (Rohani Abdul Majid et al., 2011).

Ai Ling et al. (2015) have studied the effect of kenaf loading with average size 75 μm reinforced LLDPE/polyvinyl alcohol (PVOH) on the process ability and mechanical, thermal, and water absorption. The tensile strength and elongation decreased with the increasing of kenaf fiber loading, while it is vice versa for tensile modulus. The water absorption increased as the kenaf fiber loading increased. The thermal stability of the composites increased with kenaf fiber loading; a higher temperature at the maximum weight loss rate ($T_{max}$) and char residue were observed for the composite with higher kenaf fiber loading (Ai Ling et al., 2015).

### 4.4.3 KENAF-REINFORCED POLYURETHANE COMPOSITES

Improvement of the tensile strength was observed for all prepared composites with modified short kenaf fibers (besides maleic anhydride treatment) in comparison to composites with untreated fibers. Use of maleic anhydride as a modifying agent did not give significant improvement in tensile strength. A decrease in elongation at the break was demonstrated when increasing the content of kenaf fibers, however the elongation at break values were greater after the conducted treatments. The best mechanical properties in static conditions indicated permanganate-treated, fiber-reinforced composites. It was established that a higher content of fibers in composites caused an increase in hardness and decrease in resilience values. Higher values of the mentioned properties were observed for treated fibers. The water absorption of biocomposites increased approximately two times with the incorporation of kenaf fibers. Most of treatments affected water uptake in a good way (especially for 10% of fiber loading), which means that those values are lower than for untreated fiber-reinforced composites. In the case of the higher content of fiber in composites, the greatest water uptake exhibited PU/PMn-KF 30, which can be caused due to changed and rough surface of fibers (Datta and Kopczyńska, 2015).

El-Shekeil et al. (2011) have studied the processing parameters and fiber size of kenaf-reinforced thermoplastic polyurethane composites. The method used to develop this composite consisted of two main steps: First, the influence of processing parameters such as temperature, time, and speed on tensile properties was studied. Second, the effects of different fiber sizes on tensile properties, flexural properties, and impact strength were tested. The optimum blending parameters were 190°C, 11 min, and 40 rpm for temperature, time, and speed, respectively. The tensile, flexural strength, and modulus were best for fiber size ranges between 125 and 300 µm. Impact strength showed a slight increasing trend with an increase in fiber size (El-Shekeil et al., 2011).

El-Shekeil et al. (2012) have studied the influence of fiber loading on the mechanical and thermal properties of kenaf bast fiber-reinforced thermoplastic polyurethane composites. High tensile strength and hardness observed by 30% of fiber loading, while the tensile modulus, flexural strength, and flexural modulus increased with the increase of fiber loading. In contrast, strain deteriorated with the increase of fiber content. The increase of fiber loading resulted in decline in impact strength. For thermal analysis, fiber loading decreased the thermal stability of the composite (El-Shekeil et al., 2012).

Furthermore, El-Shekeil et al. (2012) have studied the effects of alkali treatment (2%, 4%, and 6% of NaOH) on mechanical and thermal properties of kenaf fiber-reinforced thermoplastic polyurethane composites. The tensile, flexural and impact strength negatively correlates with higher concentrations of NaOH. The morphology of untreated and treated fibers showed a rougher surface in treated fibers. It also showed that some of the highest concentrations of NaOH treated fibers have NaOH residues on their surface. This was confirmed by energy dispersive X-ray point shooting performed on the same SEM machine. The morphology of the surface of fracture indicated that the untreated composite had a better adhesion. Fourier Transform Infrared (FTIR) of treated fibers showed that NaOH treatment resulted in removal of hemicelluloses and lignin. FTIR also showed that untreated composite has more H-bonding than all treated composites. Thermal characteristic studies using thermogravimetry analysis and differential scanning calorimetry showed that untreated composite was more thermally stable than treated composites (El-Shekeil et al., 2012).

Sapuan et al. (2013) studied the mechanical properties of soil buried kenaf bast fiber-reinforced thermoplastic polyurethane (TPU) composites. The reduction in the tensile properties of kenaf fiber-reinforced TPU composites in natural soil burial tests (20, 40, 60, and 80 days) dropped to 16.14MPa after 80 days of the soil burial test. The flexural strength and modulus did not show significant changes after 80 days of the soil burial test. The results indicated that TPU, as a matrix, needs more time to degrade. The low moisture content and weight gain after soil burial tests showed the poor absorbability of the composites. Weather, temperature, humidity, and degradation processes significantly influenced the reduction in tensile strength of the composites (Sapuan et al., 2013).

Ibraheem et al. (2011), was developed effective green insulation boards fabricated from polyurethane (PU) reinforced with kenaf fibers. The results show that the elastic properties increased with kenaf fiber content, and optimal performance was observed at a weight of 50% kenaf fibers. Furthermore, the minimum water absorption percentage, thickness, swelling, and change in volume were recorded at a weight of 50% kenaf fibers (Ibraheem et al., 2011).

### 4.4.4 KENAF-REINFORCED POLYLACTIC ACID (PLA) COMPOSITES

The composites with different bast short fiber loading reinforced PLA and triacetin as a plasticizer were studied by Ibrahim et al. (2009). The tensile strength and stiffness of unplasticized biocomposite materials decreased with the addition of kenaf bast fibers, but improved with the addition of triacetin. The optimum fiber loading was 30% wt of kenaf fibers in the PLA matrix with the addition of 5% triacetin. The dynamic mechanical analyses showed that triacetin improved the thermal stability of the biocomposites. The triacetin increased the storage modulus and gave a lower softening temperature for plasticized biocomposites. From scanning SEM analysis, better adhesion between the fibers and the matrix was achieved with the addition of the plasticizer (Ibrahim et al., 2009).

Shukor et al. (2014) have studied the effect of ammonium phosphate (APP) on flammability, thermal and mechanical properties of kenaf fiber-filled PLA biocomposites were investigated. APP is shown to be very effective in improving flame retardancy properties according to limiting oxygen index measurement due to increased char residue at high temperatures. However, the addition of APP decreased the compatibility between PLA and kenaf fibers resulted in significant reduction of the mechanical properties of PLA biocomposites. Thermogravimetric analysis (TGA) showed that NaOH treatment improved the thermal stability of PLA biocomposites and decreased carbonaceous char formation (Shukor et al., 2014).

Talib et al. (2011) have studied the influence of kenaf-derived cellulose (KDC) reinforced polylactic acid (PLA) composites on dynamic mechanical analysis. The storage modulus of 60% KDC/PLA composite is twice as high versus the commercial PLA and the rest of the composites within a high temperature range (above 80°C). The Tg generated from the loss modulus curves exhibit that the peak of the loss modulus was shifted to higher temperature as the percentage of the cellulose loading was increased. These results show a better thermal stability of the composites when incorporated with the cellulose (Talib et al., 2011).

Furthermore, Tawakkal et al. (2012) have studied the mechanical and physical properties of kenaf-derived cellulose-reinforced (KDC) PLA composites. The composites showed less elastic compared to neat PLA where it is around 9% of elongation. The flexural strength and modulus increased by 36% and 54%, respectively. The higher impact strength is at 30% of KDC loading in contrast the higher flexural strength is at 20% of KDC loading (Talib et al., 2011).

## REFERENCES

Ai Ling, P., Hanafi, I., & Azhar A.B. (2015). Effects of kenaf loading on processability and properties of linear low-density polyethylene/poly (vinyl alcohol)/kenaf composites. *BioResources*, 10 (4).

Aji, I.S., Sapuan, S.M., Zainudin, E.S., & Abdan K. (2009). Kenaf fibres as reinforcement for polymeric composites: A review. *International Journal of Mechanical and Materials Engineering (IJMME)*, 4 (3), 239–248.

Aji, I.S., Zainudin, E.S., khalina, A., & Khairul M.D. (2011). Studying the effect of fiber size and fiber loading on the mechanical properties of hybridized kenaf/PALF-reinforced HDPE composite. *Journal of Reinforced Plastics and Composites*, 30 (6), 546–553.

Amran, M., Izamshah, R., Hadzley, M., Shahir, M., Amri, M., Sanusi, M., & Hilmi H. (2015). *The Effect of Binder on Mechanical Properties of Kenaf Fibre/Polypropylene Composites Using Full Factorial Method.* Paper presented at the Applied Mechanics and Materials.

Anuar, H., & Zuraida A. (2011). Improvement in mechanical properties of reinforced thermoplastic elastomer composite with kenaf bast fibre. *Composites Part B: Engineering*, 42 (3), 462–465.

Asumani, O.M.L., Reid, R.G., & Paskaramoorthy R. (2012). The effects of alkali–silane treatment on the tensile and flexural properties of short fibre non-woven kenaf reinforced polypropylene composites. *Composites Part A: Applied Science and Manufacturing*, 43 (9), 1431–1440.

Azwa, Z.N., & Yousif B.F. (2013). Characteristics of kenaf fibre/epoxy composites subjected to thermal degradation. *Polymer Degradation and Stability*, 98 (12), 2752–2759.

Bachtiar, D.S., S.M., Hamdan, M.M. (2008). The effect of alkaline treatment on tensile properties of sugar palm fibre reinforced epoxy composites. *Materials & Design*, 29 (7), 1285–1290.

Batouli, S.M., Zhu, Y., Nar, M., & D'Souza N.A. (2014). Environmental performance of kenaf-fiber reinforced polyurethane: a life cycle assessment approach. *Journal of Cleaner Production*, 66, 164–173.

Bernard, M., Khalina, A., Ali, A., Janius, R., Faizal, M., Hasnah, K.S., & Sanuddin A.B. (2011). The effect of processing parameters on the mechanical properties of kenaf fibre plastic composite. *Materials & Design*, 32 (2), 1039–1043.

Bos, H. (2004). *Phd thesis, The potential of flax fibers as reinforcement for composite materials.* PhD, Technische Universiteit Eindhoven, Netherland.

Callister, W.D.R., D.G. (2012). *Fundamentals of Materials Science and Engineering: An Integrated Approach*: Wiley.

Chin, C.W., & Yousif B.F. (2009). Potential of kenaf fibres as reinforcement for tribological applications. *Wear*, 267 (9–10), 1550–1557.

Datta, J., & Kopczyńska P. (2015). Effect of kenaf fibre modification on morphology and mechanical properties of thermoplastic polyurethane materials. *Industrial Crops and Products*, 74, 566–576.

Deka, H., Misra, M., & Mohanty A. (2013). Renewable resource based "all green composites" from kenaf biofiber and poly(furfuryl alcohol) bioresin. *Industrial Crops and Products*, 41, 94–101.

El-Shekeil, Y., Sapuan, S., Khalina, A., Zainudin, E., & Al-Shuja'a O. (2012). Influence of chemical treatment on the tensile properties of kenaf fiber reinforced thermoplastic polyurethane composite. *eXPRESS Polymer Letters*, 6 (12).

El-Shekeil, Y.A., Salit, M.S., Abdan, K., & Zainudin E.S. (2011). Development of a new kenaf bast fiber-reinforced thermoplastic polyurethane composite. *BioResources*, 6 (4), 4662–4672.

El-Shekeil, Y.A., Sapuan, S.M., Abdan, K., & Zainudin E.S. (2012). Influence of fiber content on the mechanical and thermal properties of Kenaf fiber reinforced thermoplastic polyurethane composites. *Materials & Design*, 40, 299–303.

El-Shekeil, Y.A., Sapuan, S. M., Khalina, A., Zainudin, E.S., & Al-Shuja'a O.M. (2012). Effect of alkali treatment on mechanical and thermal properties of Kenaf fiber-reinforced thermoplastic polyurethane composite. *Journal of Thermal Analysis and Calorimetry*, 109 (3), 1435–1443.

Faruk, O.B., A.K. Fink, & H.P. Sain M. (2012). Biocomposites reinforced with natural fibers: 2000–2010. *Progress in Polymer Science*, 37 (11), 1552–1596.

Fu, S.-Y., Feng, X.-Q., Lauke, B., & Mai Y.-W. (2008). Effects of particle size, particle/matrix interface adhesion and particle loading on mechanical properties of particulate–polymer composites. *Composites Part B: Engineering*, 39 (6), 933–961.

Hao, A., Zhao, H., & Chen J.Y. (2013). Kenaf/polypropylene nonwoven composites: The influence of manufacturing conditions on mechanical, thermal, and acoustical performance. *Composites Part B: Engineering*, 54, 44–51.

Hao, A., Zhao, H., Jiang, W., Yuan, L., & Chen J.Y. (2012). Mechanical properties of kenaf/polypropylene nonwoven composites. *Journal of Polymers and the Environment*, 20 (4), 959–966.

Holbery, J., & Houston D. (2006). Natural-fiber-reinforced polymer composites in automotive applications. *Jom*, 58 (11), 80–86.

Ibraheem, S.A., Ali, A., & Khalina A. (2011). Development of green insulation boards from kenaf fibres and polyurethane. *Polymer-Plastics Technology and Engineering*, 50 (6), 613–621.

Ibrahim, N.A., Yunus, W.M.Z.W., & Khalina A. (2009). Poly(Lactic Acid) (PLA)-reinforced kenaf bast fibers composites: The effect of triacetin. *Journal of Reinforced Plastics and Composites*.

Ishak, M.R., Sapuan, S.M., Leman, Z., Rahman, M.Z.A., Anwar, U.M.K., & Siregar J.P. (2013). Sugar palm (Arenga pinnata): Its fibres, polymers and composites. *Carbohydrate Polymers*, 91 (2), 699–710.

John, M.J., Bellmann, C., & Anandjiwala R.D. (2010). Kenaf–polypropylene composites: Effect of amphiphilic coupling agent on surface properties of fibres and composites. *Carbohydrate Polymers*, 82 (3), 549–554.

Kamaruddin, S., Abdan, K., Ali, A., Maringgal, B., Jamaliah, S., & Yunus W.M.Z.M. (2011). Kenaf performance in PP/EVA/clay biocomposite. *Materials Testing*, 53 (6), 364–368.

Kestur, G.S.G., & G.C.A. Fernando, W. (2009). Biodegradable composites based on lignocellulosic fibers—An overview. *Progress in Polymer Science*, 34 (9), 982–1021.

Kim, K.-J. (2015). Modification of nano-kenaf surface with maleic anhydride grafted polypropylene upon improved mechanical properties of polypropylene composite. *Composite Interfaces*, 22 (6), 433–445.

Koronis, G.S., & A. Fontul, M. (2013). Green composites: A review of adequate materials for automotive applications. *Composites Part B: Engineering*, 44 (1), 120–127.

Ku, H.W., H. Pattarachaiyakoop, & N. Trada M. (2011). A review on the tensile properties of natural fiber reinforced polymer composites. *Composites Part B: Engineering*, 42 (4), 856–873.

Kwon, H.-J., Sunthornvarabhas, J., Park, J.-W., Lee, J.-H., Kim, H.-J., Piyachomkwan, K., Cho, D. (2014). Tensile properties of kenaf fiber and corn husk flour reinforced poly (lactic acid) hybrid bio-composites: Role of aspect ratio of natural fibers. *Composites Part B: Engineering*, 56, 232–237.

La Mantia, F.P., & Morreale M. (2011). Green composites: A brief review. *Composites Part A: Applied Science and Manufacturing*, 42 (6), 579–588.

Leman, Z.S., Saifol, S.M., Maleque, A.M., Ahmad, M.A., & M.M.H.M. (2008). Moisture absorption behavior of sugar palm fiber reinforced epoxy composites. *Materials & Design*, 29 (8), 1666–1670.

Li, X., Tabil, L.G., & Panigrahi S. (2007). Chemical treatments of natural fiber for use in natural fiber-reinforced composites: A review. [journal article] *Journal of Polymers and the Environment*, 15 (1), 25–33.

Liu, W.M., Askeland, A.K., Drzal, P., Misra, L.T., M. (2004). Influence of fiber surface treatment on properties of Indian grass fiber reinforced soy protein based biocomposites. *Polymer*, 45 (22), 7589–7596.

Maringgal, B., Abdau, K., Ali, A., Janius, R., Faizal, M., & Kamaruddin S. (2011). Optimization of compression moulding temperature for polypropylene materials. *Materials Testing*, 53 (5), 280–284.

Mazumdar, S. (2001). *Composites Manufacturing: Materials, Product, and Process Engineering*: CRC Press.

Md Radzi, M.K.F., Sulong, A.B., Muhamad, N., Mohd Latiff, M.A., & Ismail N.F. (2015). Effect of filler loading and NaOH addition on mechanical properties of moulded kenaf/polypropylene composite *Pertanika J. Trop. Agric. Sci.*, 38 (4), 583–590.

Mohanty, A.K., Misra, M., & Drzal L.T. (2001). Surface modifications of natural fibers and performance of the resulting biocomposites: An overview. *Composite Interfaces*, 8 (5), 313–343.

Mohanty, A.K.M., M. Drzal, L.T. (2005). *Natural Fibers, Biopolymers, and Biocomposites*: Taylor & Francis.

Mohd Nurazzi, N., Khalina, A., Sapuan, S.M., Dayang Laila, A.H.A.M., Rahmah, M., & Hanafee Z. (2017). A review: Fibres, polymer matrices and composites. *Pertanika J. Sci. & Technol.*, 25 (4), 1085–1102.

Narish, S., Yousif, B.F., & Rilling D. (2011). Adhesive Wear of Thermoplastic Composite Based on Kenaf Fibres. *Journal of Engineering Tribology*, 225 (2), 101–109.

Nor Azowa, Ibrahim, Kamarul, Arifin Hadithon, & Abdan K. (2009). Effect of Fiber Treatment on Mechanical Properties of Kenaf Fiber-Ecoflex Composites. *Journal of Reinforced Plastics and Composites*.

Norlin, N., Akil, H.M., Mohd Ishak, Z.A., & Abu Bakar A. (2011). Behavior of kenaf fibers after immersion in several water conditions. *Bioresources.com*, 6 (2), 950–960.

Nunna, S., Chandra, P.R., Shrivastava, S., & Jalan A. (2012). A review on mechanical behavior of natural fiber based hybrid composites. *Journal of Reinforced Plastics and Composites*, 31 (11), 759–769.

Ochi, S. (2008). Mechanical properties of kenaf fibers and kenaf/PLA composites. *Mechanics of Materials*, 40 (4–5), 446–452.

Paridah, M.T., Basher, A.B., SaifulAzry, S., & Ahmed Z. (2011). Retting process of some bast plant fibres and its effect on fibre quality: A review. *BioResources*, 6 (4), 5260–5281.

Paukszta, D., & Borysiak S. (2013). The influence of processing and the polymorphism of lignocellulosic fillers on the structure and properties of composite materials—A review. *Materials*, 6 (7), 2747–2767.

Pereira, P.H.F., Rosa, M.d.F., Cioffi, M.O.H., Benini, K.C.C.d.C., Milanese, A.C., Voorwald, H.J.C., & Mulinari D.R. (2015). Vegetal fibers in polymeric composites: A review. *Polímeros*, 25, 9–22.

Rassiah, K., & Megat Ahmad M.M.H. (2013). A review on mechanical properties of bamboo fiber reinforced polymer composite. *Australian Journal of Basic and Applied Sciences*, 7 (8), 247–253.

Rohani Abdul Majid, Hanafi Ismail, & Taib R.M. (2011). Effects of polyethylene-g-maleic anhydride on properties of low density polyethylene/thermoplastic sago starch reinforced kenaf fibre composites. *Iranian Polymer Journal*, 19 (7), 501–510.

Rowell, R.M., Sanadi, A., Jacobson, R., & Caulfield D.F. (1999). Properties of kenaf/polypropylene composites. In J. Terry Sellers & A. R. Nancy (Eds.), *Kenaf Properties, Processing and Products*, (pp. 381–392). Mississippi State University, Ag & Bio Engineering Box 9632, Mississippi State, MS 39762: Treesearch.

Saba, N., Paridah, M., & Jawaid M. (2015). Mechanical properties of kenaf fibre reinforced polymer composite: A review. *Construction and Building Materials*, 76, 87–96.

Salleh, F.M., Hassan, A., Yahya, R., & Azzahari A.D. (2014). Effects of extrusion temperature on the rheological, dynamic mechanical and tensile properties of kenaf fiber/HDPE composites. *Composites Part B: Engineering*, 58, 259–266.

Sapuan, S.M., Pua, F.-l., El-Shekeil, Y.A., & Al-Oqla F.M. (2013). Mechanical properties of soil buried kenaf fibre reinforced thermoplastic polyurethane composites. *Materials & Design*, 50, 467–470.

Shalwan, A., & Yousif B.F. (2013). In State of Art: Mechanical and tribological behaviour of polymeric composites based on natural fibres. *Materials & Design*, 48, 14–24.

Shukor, F., Hassan, A., Saiful Islam, M., Mokhtar, M., & Hasan M. (2014). Effect of ammonium polyphosphate on flame retardancy, thermal stability and mechanical properties of alkali treated kenaf fiber filled PLA biocomposites. *Materials & Design*, 54, 425–429.

Smith, W.F., & Hashemi J. (2006). *Foundations of Materials Science and Engineering* (4th ed.): McGraw-Hill.

Sreekala, M.S.K., M.G. Joseph, S. Jacob, M. Thomas, S. (2000). Oil Palm Fibre Reinforced Phenol Formaldehyde Composites: Influence of Fibre Surface Modifications on the Mechanical Performance. *Applied Composite Materials*, 7 (5–6), 295–329.

Subasinghe, A.D.L., Das, R., & Bhattacharyya D. (2015). Fiber dispersion during compounding/injection molding of PP/kenaf composites: Flammability and mechanical properties. *Materials & Design*, 86, 500–507.

Suharty, N.S., Almanar, I.P., Sudirman, Dihardjo, K., & Astasari N. (2012). Flammability, Biodegradability and Mechanical Properties of Bio-Composites Waste Polypropylene/Kenaf Fiber Containing Nano CaCO3 with Diammonium Phosphate. *Procedia Chemistry*, 4, 282–287.

Talib, R.A., Tawakkal, I.S.M.A., & Khalina A. (2011). The Influence of Mercerised Kenaf Fibres Reinforced Polylactic Acid Composites on Dynamic Mechanical Analysis. *Key Engineering Materials*, 471–472, 815–820.

Tawakkal, I.S.M.A., Talib, R.A., Abdan, K., & Ling C.N. (2012). Mechanical and physical properties of kenaf-derived cellulose (KDC)-filled polylactic acid (PLA) composites. *BioResources*, 7 (2), 1643–1655.

Tee, Y.B., Rosnita, A.T., Khalina, A., Chin, N.L., Roseliza, K.B., & Khairul Faezah M.Y. (2015). Reinforcing Mechanical, Water Absorption and Barrier Properties of Poly(Lactic Acid) Composites with Kenaf-Derived Cellulose of Thermally-Grafted Aminosilane. *Pertanika J. Trop. Agric. Sci*, 38 (4), 563–573.

Thiruchitrambalam, M., Alavudeen, A., Athijayamani, A., Venkateshwaran, N., & Perumal A.E. (2009). Improving mechanical properties of banana/kenaf polyester hybrid composites using sodium laulryl sulfate treatment. *Mater Phys Mech*, 8, 165–173.

# 5 Effect of Silica Aerogel on Polypropylene Reinforced with Kenaf Core Fiber for Interior Automotive Components

*A.S. Harmaen, M.T. Paridah, M. Jawaid, A.M. Fariz, and B. Asmawi*

## CONTENTS

5.1 Introduction ..................................................................................................81
5.2 Materials ......................................................................................................85
5.3 Methods ........................................................................................................85
    5.3.1 Kenaf Core Processing ....................................................................85
    5.3.2 Kenaf Board Fibricating ..................................................................85
    5.3.3 Mechanical Testing .........................................................................86
5.4 Results and Discussion .................................................................................87
    5.4.1 Mechanical Properties ....................................................................87
5.5 Conclusion ...................................................................................................90
Acknowledgment ..................................................................................................91
References .............................................................................................................91

## 5.1 INTRODUCTION

The demand from automotive companies for materials with noise abatement capabilities as well as increased fuel efficiency by weight reduction has increased (Wambua et al. 2003) due to the fact that natural fiber composites possess excellent sound-absorbing capabilities, are more shatter resistant, and have more efficient energy management characteristics than glass, therefore the demand for natural fiber composites has increased in the market (Wegst 1996) (Figure 5.1). The demand for natural fibers in plastic composites are forecast to grow at a rate of 15–20% in automobile applications and 50% or more in selected building application (Ticolau et al. 2010). Natural fiber-based automobile parts such as various panels, trim parts, and

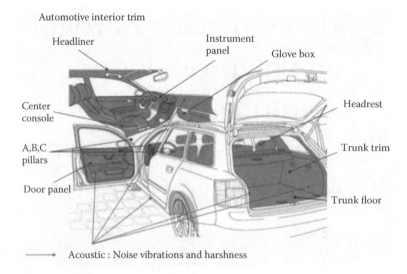

**FIGURE 5.1** Automotive parts.

brake shoes are attractive to the automotive industry because they have reduced the weight of parts by more than 10% and have also brought the cost down by as much as 5% (Karus et al. 2000) (Figure 5.1). At present, conventional plastic materials such as polypropylene (PP) play an essential role in consumers' lives due to their excellent properties.

PP exhibits good strength, is lightweight, is easily processed, and is economical as well (Figure 5.2). However, the widespread consumption of plastics in the packaging,

**FIGURE 5.2** Polypropylene.

agricultural, and automobile industries has many implications for the environment. PP consists of homopolymers and copolymers. PP is common melting temperature in range at 161–165°C. However, copolymers contain ethylene commoners and are sequentially are subdivided to random and block copolymers. The melting temperature for copolymers is 140–155°C and the specific gravity (density) of polypropylene is about 0.90–0.91 g/cm$^3$ (Anatole 2007). Polypropylene is used in different applications such as packaging, transportation, and textiles. According to previous study, PP was widely used in automotive components (Kozlowski 2012).

Silica aerogel has unusual solid material properties that have a high potential in various applications. The size of pores depends on depending on the purity and the fabrication method. For pure aerogel, the average pore diameter is between 10 and 100 nm (Zeng et al. 1994), but generally pore sizes are between 5 and 70 nm (Figure 5.3) (van Bommel et al. 1997). The extraordinary small pore sizes and high porosity of aerogel gives remarkable physical, thermal, optical, and acoustical properties, but a very low mechanical strength.

Silica aerogel has a skeleton density of aerogel is around 2200 kg/m$^3$, but the high porosity structure gives a bulk density as low as 3 kg/m$^3$ (Hunt et al. 1991). However, aerogels produce for building applications have an overall density of 70–150 kg/m$^3$. Although silica aerogels have considerable high compression strength up to 3 bar, they have a very low tensile strength makes the material very fragile. Aerogel need to be well hydrophobised, to prevent water contact that could demolish the aerogel structure due to surface tension in the pores (Jensen et al. 2004). Therefore, in most applications, aerogel is often used in combination with a vacuum to prevent water inclusion and thermal conductivity. However, for commercial aerogel insulation materials, the weak tensile properties problems are solved by incorporating them into a fiber matrix.

Kenaf (*Hibiscus cannabinus* L., family Malvaceae) is an herbaceous annual plant that can be grown in various temperate conditions. Depending on variety and

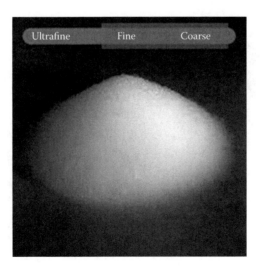

**FIGURE 5.3** Nano aerogel.

agronomy of planting, it can grow larger than 3 m within 3 months with stem diameter of 25–51 mm (Figures 5.4 and 5.5). As a dicotyledonous plant, the kenaf stalk has three layers; an outer cortical layer known as the "bast," an inner woody layer known as the "core," and a thin sponge-like tissue layer with mostly nonferrous cells at the center known as the "pith" (Ashori et al. 2006). Kenaf been used through the ages in making ropes, canvas, and sacks.

**FIGURE 5.4** Kenaf tree plantations.

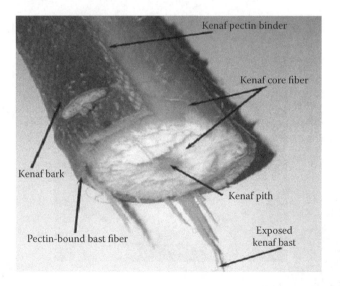

**FIGURE 5.5** Stem of kenaf.

The new interest in kenaf cultivation have been achieved for two main reasons; kenaf's ability to absorb nitrogen and phosphorus in soil and its ability to accumulate carbon dioxide at a high rate (Abe and Ozaki, 1998; Amaducci et al. 2000). The merits of using natural lignocellulosic fibres as reinforcements of the matrix in composites creates a new oppurtinity for kenaf usage. Nishino et al. (2003) reported that kenaf exhibits low density, non-abrasiveness during processing, high specific mechanical properties, and biodegradability. Apart from that, kenaf is used as an alternative to wood in pulp and paper industries to reduce deforestation (Pande and Roy, 1998), as non-woven mats in the automotive industries (Magurno 1999), and in textiles (Ramaswamy et al. 1995) and fibreboard (Kawai et al. 2000). The kenaf plant is clearly a cellulosic source with many economic and ecological advantages.

The automotive interior manufacture is a thrust area for the biomass R&D. Research has demonstrated that the use of natural vegetable plant fibers such as kenaf, ramie, flax, hemp, and cotton for automotive composite applications has many advantages (Mueller et al., 2001; Mueller and Stryjewski, 2001). From a technical point of view, these biobased composites will enhance mechanical strength and acoustic performance, reduce material weight and processing time, lower production cost, improve passenger safety and shatterproof performance under extreme temperature changes, and improve biodegradability for the auto interior parts. The conversion of these agricultural crops into automotive interior materials will benefit the establishment of a sustainable and environmentally friendly resource base for the industry.

The aim of this study is to investigate the mechanical properties of silica aerogel upon the incorporation of PP and kenaf core fiber composites.

## 5.2 MATERIALS

PP pellets were supplied by Chemical Titan Holding Sdn. Bhd., Pasir Gudang, Johor. Kenaf fibers were supplied by National Kenaf and Tabacco Board (LKTN), Kota Baharu, Kelantan. Silica aerogel was supplied by Maerotech Sdn. Bhd. Nilai, Negeri Sembilan, Malaysia.

## 5.3 METHODS

### 5.3.1 KENAF CORE PROCESSING

After 4 or 5 months in plantation, the kenaf tree was cut down, then the stems and leaves were separated (Figure 5.6). The kenaf stems were chipped at the laboratory using a chipper machine.

The kenaf chips were then grained into small particles 0.5–1.0 mm in size. The kenaf core fibers should be stored in a tight plastic bag before usage.

### 5.3.2 KENAF BOARD FIBRICATING

Figure 5.7 shows the kenaf board fibricating process. Polypropelene and kenaf core fibers were first dried in a vacuum oven at 80°C for 24 h to remove water.

**FIGURE 5.6** Kenaf core processing. (a) Kenaf plantation, (b) harvesting, (c) chipping, (d) kenaf chip, and (e) kenaf core fibers.

The composites of PP/AG/KCF were compounded in a twin screw mixer at 175°C at a speed of 50 rpm (Table 5.1).

After being compounded, the samples were crushed into small pellet size. The weight ratios of AG in PP/KCF were 0, 1%, 3%, and 5%. The temperature of the mold was 175°C for 9 min, and then pressed and cooled for about 3 min. The specimens were left at room temperature at 23°C for 24 h before testing.

### 5.3.3 Mechanical Testing

The samples for testing tensile and impact strength were prepared using a hot compression molding machine. A mold with the specific dimensions of 150 × 150 × 1 mm and 3 mm. (length × width × thickness) was used for the sample preparation. For PP/AG/KCF composites, the molding cycle involved 3 min for preheating, 7 min for compression, and 3 min for cooling. Tensile tests were performed using a universal testing machine (Instron-3366) at a crosshead speed of 5 mm/min. The tensile strength test was performed according to ASTM D638-10. The impact izod test was performed using impactor machine, according ASTM Standard D777. For every set of formulations, 5 specimens were tested to determine the average properties. Prior to testing, the specimens were conditioned at room temperature in a desiccator for 24 h. The tests were conducted at a standard laboratory atmosphere of 23°C (±2°C).

# Effect of Aerogel on Polyprolene with Kenaf Core Fiber

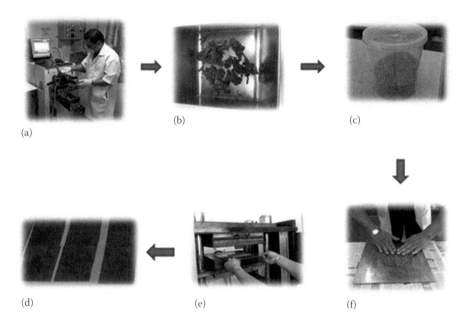

**FIGURE 5.7** Kenaf board fibricating. (a) The compounding using internal mixer (brabender) machine, (b) pellet compounding process, (c) small pellets, (d) kenaf core fiber composite, (e) hot and cold pressing, and (f) Kenaf-PP pellets placed into a mold.

**TABLE 5.1**
**Composition of PP, Silica Aerogel, and KCF Composites Ratio**

| Number of Samples | PP | AG | KCF |
|---|---|---|---|
| PP | 40 g | – | – |
| PP/AG1% | 39 g | 1 g | – |
| PP/AG3% | 37 g | 3 g | – |
| PP/AG 5% | 35 g | 5 g | – |
| PP/AG1%/KCF50% | 19 g | 1 g | 20 g |
| PP/AG3%/KCF50% | 17 g | 3 g | 20 g |
| PP/AG5%/KCF50% | 15 g | 5 g | 20 g |

*Note:* Kenaf core fiber (KCF), Polypropylene (PP), Silica aerogel (AG)

## 5.4 RESULTS AND DISCUSSION

### 5.4.1 Mechanical Properties

The mechanical improvement of a fiber-reinforced plastic composite mostly depends on three main factors: the modulus and strength of the fiber, the chemical stability and strength of the resin, and the effectiveness of the bond between resin and fiber

## TABLE 5.2
### The Mechanical Tests of the Composites in Comparison to Silica Aerogel Ratios

| Board Type | Composite | Density (g/cm³) | Tensile strength (MPa) | Tensile modulus (Mpa) | Izod impact (Kj/m²) |
|---|---|---|---|---|---|
| Pure PP | PP | 0.79 | 31.2[a] | 747[d] | 3.4[a] |
| PPC | PP/AG 1% | 0.80 | 28.9[b] | 716[d] | 2.6[b] |
|  | PP/AG 3% | 0.80 | 31.0[a] | 756[d] | 2.3[c] |
|  | PP/AG 5% | 0.81 | 30.1[a] | 751[d] | 1.8[d] |
| KPPC | PP/KCF | 0.98 | 14.1[d] | 1322[c] | 1.5[e] |
|  | PP/KCF/AG1% | 1.00 | 15.0[d] | 1574[b] | 1.4[e] |
|  | PP/KCF/AG3% | 1.04 | 17.4[c] | 1866[a] | 1.5[e] |
|  | PP/KCF/AG5% | 1.04 | 18.1[c] | 1772[a] | 1.5[e] |

*Note:* Means followed by the same letter [a, b, c, d] in same column were not significantly different at $p \leq 0.05$; polypropylene (PP), silica aerogel (AG), kenaf core fiber (KCF).

(Joseph et al. 1996). Table 5.2 shows the tensile strength, tensile modulus, izod impact strength, and density values of the composites.

Figure 5.8 illustrates the effects of Silica Aerogel ratios on the tensile strength of composites. A tensile test is to measures the resistance of a composite when load is applied with the elongation of the specimen over some distance. The pure PP composite exhibited the highest tensile strength with 31.2 Mpa. Thus, the PP with AG 3% and 5% composites has the same group of strength with pure PP. According to Osswald (1999), stress is capably transferred when the bond between fiber and matrix is good. At the critical fiber length, the stress is transfused from the matrix to fiber, resulting in a stronger composite. The addition of KCF with PP was significantly decreased the tensile strength of the composites. It showed that PP/KCF composite had tensile strength with 14.1 Mpa and placed at the lowest group. In fact, by presence of AG with KCF and PP of the composite increased the tensile strength significantly. It showed that PP/KCF/AG 5% composite had the tensile strength of 18.1 Mpa.

Figure 5.9 illustrates the effects of AG ratios on the tensile modulus of composites. The presence of AG with PP of the composite did not increase the tensile modulus. It showed that pure PP composite with tensile modulus of 747 MPa was in the same group with PP/AG ratios (1%, 3%, and 5%), and the composite and had the lowest tensile modulus. According to previous studies, the merger of cellulosic or fiber had improved the stiffness of the composite materials (Ismail et al. 2011). Thus, KCF was found to increase the tensile modulus of kenaf polypropylene polymer composite (KPPC). This suggests that the incorporation of KCF and AG in this process significantly increases the tensile strength. The PP/KCF/AG3% composite with 1866 Mpa had a highest tensile strength compared with PP/KC composite with 1332 Mpa. Between KCF and AG, KCF is more influencing in tensile modulus. However, by addition of AG, the tensile modulus of KPPC was increased significantly.

# Effect of Aerogel on Polyprolene with Kenaf Core Fiber

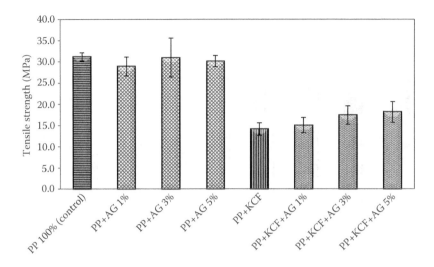

**FIGURE 5.8** The tensile strength values of the different types of composites in comparison to silica aerogel ratios.

Chemical composition plays important role in mechanical properties of plant's fiber. Habibi et al. (2008), mentioned that cellulose, hemicellulose, and lignin contents were found to have a strong influence on the mechanical properties of the fiber itself. Among these three components, the cellulose content is the most influential factor that contributed to mechanical properties of the fibers. This is because cellulose

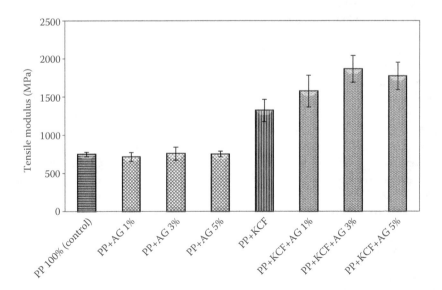

**FIGURE 5.9** The tensile modulus values of the different types of composites in comparison to silica aerogel ratios.

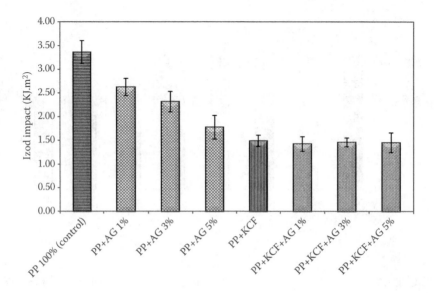

**FIGURE 5.10** The izod impact values of the different types of composites in comparison to silica aerogel ratios.

is one of the stiffest and strongest organic constituents in a natural fiber and it is the the main structural component that provides strength and stability to the plant cell walls and the fibers (Ray and Rout 2005). Generally, it is well accepted that the higher cellulose content leads to higher mechanical properties of the fibers. For kenaf fibers, the highest mechanical properties of kenaf bast fibers is expected compare to core fibers, as the cellulose content for the bast and core fibers are 60.8% and 50.6%, respectively.

Figure 5.10 illustrates the effects of AG ratios on the izod impact strength of the composites. According to Nielson and Landel (1994), to enhance the impact strength, the fiber particles must have a larger ductility than the matrix. The properties of pure PP as a polymer without any addition of fiber showed the highest impact strength value with 3.4 kJ/m$^2$. According to Lee et al. (2013), PP exhibits a light weight, good strength, and is easy to process due its excellent properties. The addition of AG ratios (1%, 3%, and 5%) with PP significantly decreasing the impact strength of the composites from 3.4 kJ/m$^2$ to 1.8 kJ/m$^2$. Meanwhile, the addition of KCF was also giving the same decreasing result. The PP/KCF composite impact strength was 1.5 kJ/m$^2$, and by adding the AG ratios (1%, 3%, and 5%), it still had not improved the impact strength of the composite.

## 5.5 CONCLUSION

The study has shown that kenaf core can be mixed with polypropylene with the presence of silica aerogel to produce composites of acceptable mechanical properties. Both the impact and tensile strength were markedly reduced with the addition of KCF and AG. However, the presence of AG kept this reduction minimal. The study has shown that the addition of AG was able to improve all the effects of kenaf core fibers: improve tensile strength and increase tensile modulus. The addition of AG ratios (1%, 3%, and 5%) with PP significantly decreased the impact strength of the composites.

## ACKNOWLEDGMENT

We like to thank Maerotech Sdn. Bhd. to supplied silica aerogel material during study.

## REFERENCES

Abe, K., and Ozaki, Y. 1998. Comparison of useful terrestrial and aquatic plant species for removal of nitrogen and phosphorus from domestic waste water. *Soil Science and Plant Nutrition* (Japan).

Amaducci, S., Amaducci M.T., Benati R., and Venturi, G. 2000. Crop yield and quality parameters of four annual fibre crops (hemp, kenaf, maize and sorghum) in the North of Italy. *Industrial Crops & Products* 11: 179–186.

Anatole, A.K. 2007. *Wood-plastic Composites*. John Wiley & Sons, USA.

Ashori, A., Harun J., Raverty W., and Yusoff, M. 2006. Chemical and morphological characteristics of Malaysian cultivated kenaf (*Hibiscus cannabinus*) fiber. *Plastics Technology and Engineering* 45: 131–134.

Habibi, Y., El-Zawawy, W., Ibrahim, M.M., and Dufresne, A. 2008. Processing and characterization of reinforced polyethylene composites made with lignocellulosic fibres from Egyptian agro-industrial residues. *Journal of Composites Science and Technnology* 68: 1877–1885.

Hunt, A.J., Jantzen, C.A., and Cao, W. 1991. Aerogel a high performance insulating material at 0.1 bar, in: R.S. Graves, D.C. Wysocki (Eds.), ASTM STP 1116, *Insulation Materials: Testing and Application, Proceedings of the Symposia for Committee 16 on Insulation Materials*, Gatlinburg, October 10–12, vol. 2, 455–463.

Ismail, H., Abdullah, A.H., and Bakar, A.A. 2011. Influence of acetylation on the tensile properties, water absorption, and thermal stability of high-density polyethylene, soya powder and kenaf core composites. *Journal of Vinyl and Additive Technology* 17: 132–137.

Ismawati, P. 2006. Carboxymethylation of cellulose from kenaf (*Hibiscus cannabinus* L.) core for hydrogel production. Master's Thesis Universiti Putra Malaysia.

Jensen, K.I., Schultz, J.M., and Kristiansen, F.H. 2004. Development of windows based on highly insulating aerogel glazings. *Journal of Non-Crystalline Solids* 350: 351–357.

Joseph, K., Varghese, S., Kalaprasad, G., Thomas, S., Prasannakumari, L., Koshy, P., and Pavithran, C. 1996. Influence of interfacial adhesion on the mechanical properties and fracture behaviour of short sisal fibre reinforced polymer composites. *European Polymer Journal* 32: 1243–1250.

Karus, M., Kamp, M., and Lohmeyer, D. 2000. *Study of markets and price situation of natural fibres (Germany and EU)*, Nova Institute, Germany.

Kawai, S., Ohnishi Y., Okudaira Y., and Zhang, M. 2000. Manufacture of oriented fiberboard from kenaf bast fibers and its application to the composite panel. *Intern. Kenaf Symp.* 144–148.

Khalil A.H.P.S., Chow, W.C., Rozman, H.D., Ismail, H., Ahmad, M.N., and Kumar, R.N. 2001. The effect of anhydride modification of sago starch on the tensile and water absorption properties of sago filled linear low density polyethylene (LLDPE). *Polymer Plastic Technology Engineering* 40: 249–263.

Kozlowski, R., Muzyczek, M., and Mieleniak, B. 2008. Upholstery fire barriers based on natural Fibers. *Journal of Natural Fibers* 1: 85–95.

Kozlowski, M. 2012. *Lightweight plastic materials*. ISBN: 978-953-51-0346-2, *InTech*.

Lee, J.M., Mohd, I.Z.A., Mat, T.R., Law, T.T., and Ahmad, T.M.Z. 2013. Mechanical, thermal and water absorption properties of kenaf fiber based polypropylene and poly (butylene succinate) composites. *Journal of Polymer and the Environment* 21: 293–302.

Magurno, A. 1999. Vegetable Fiber in automotive interior components. Die Angewandte Makromolekulare Chemie, 272(1): 99–107.

Mueller D.H., Krobjilowski A., and Muessig, J. 2001. Proceedings of Beltwide Cotton Conferences, National Cotton Council of America, Memphis, TN. 689–696.

Mueller D.H., and Stryjewski D.D. 2001. Proceedings of Beltwide Cotton Conferences, National Cotton Council of America, Memphis, TN. 714–717.

National Kenaf & Tobacco Board (LKTN). 2013. *Kenaf basic information* (2nd edition).

Nielson, L.E. and Landel R.F. 1994. *Mechanical Properties of Polymers and Composites*. New York: Marcel Dekker, Inc.

Nishino, T., Hiroa, A.K., Kotera, M., Nakamae K., and Inagaki, H. 2003. Kenaf reinforced biodegradable composite. *Composites Science and Technology* 63: 1281–1286.

Osswald, T.A. 1999. Fundamental principles of polymer composites: Processing and design. In: Proceedings of 5th International Conference of Wood Fiber Plastic Composites. Madison, Wisconsin, May 26–27.

Pande, H., and Roy, D.N. 1998. Influence of fibre morphology and chemical composition on the papermaking potential of kenaf fibres. Pulp & paper Canada.

Ramaswamy, G.N., Craft, S., and Wartelle, L. 1995. Uniformity and softness of kenaf fibers for textileproducts. *Textile Research Journal* 65: 765–769.

Ray, D. and Rout J. 2005. Thermoset biocomposites. *Natural Fibers, Biopolymers, and Biocomposites*. 291–345, Boca Raton: CRC Press.

Reddy, N. and Yang, Y. 2005. Biofibres from agricultural byproducts for industrial applications *Journal of Trends in Biotechnology* 23: 1–6.

Sellers Jr., T., Reichert N.A., Columbus E., Fuller M., and Williams, K. 1999. *Kenaf Properties, Processing and Products*. Mississippi State University, MS.

Ticolau, A., Aravinthan, T., and Cardona, F. 2010. A Review of current development in natural fiber composites for structural and infrastructure applications, University of Southern Queensland, Toowoomba, SREC2010-F1-5.

Van Bommel, M.J., den Engelsen, C.W., and Van Miltenburg, J.C. 1997. A thermoporometry study of fumed silica/aerogel composites. *Journal of Porous Materials* 4: 143–150.

Wambua, P., Ivens, J., and Verpoest, I. 2003. Natural fibres: Can they replace glass in fibre reinforced plastics? *Composite Science and Technology* 63: 1259–1264.

Wegst, U.G.K. 1996, PhD Thesis, University of Cambridge, UK.

Zeng, S.Q., Hunt, A.J., Cao, W., and Greif, R. 1994. Pore size distribution and apparent thermal conductivity of silica aerogel. *Journal of Heat Transfer* 116: 756–759.

# 6 Impact of Silane Treatment on the Properties of Kenaf Fiber Unsaturated Polyester Composites

*Md. Rezaur Rahman, Sinin Hamdan, and Rubiyah bt Hj Baini*

**CONTENTS**

| | | |
|---|---|---|
| 6.1 | Introduction | 94 |
| 6.2 | Natural Fiber Composites | 95 |
| | 6.2.1 Overview of the Composites | 95 |
| | 6.2.2 Fiber Composites | 95 |
| | 6.2.3 Benefits of Natural Fiber-Reinforced Composites | 96 |
| | 6.2.4 Drawbacks of Natural Fiber-Reinforced Composites | 96 |
| | 6.2.5 Trends of Fiber Reinforced | 97 |
| | 6.2.6 Biocomposites | 97 |
| | 6.2.7 Hybrid Composites | 98 |
| | 6.2.8 Advantages of Natural Fibers in Composites | 98 |
| | 6.2.9 Applications of Natural Fibers | 98 |
| 6.3 | Fibers | 98 |
| | 6.3.1 Kenaf | 98 |
| | 6.3.2 Properties of Kenaf Fibers | 99 |
| | 6.3.3 Advantages of Kenaf | 99 |
| | 6.3.4 Trend in Kenaf Composites | 100 |
| | 6.3.5 The Drawback of Kenaf to Its Composites | 102 |
| 6.4 | Chemical Treatment | 102 |
| | 6.4.1 Pretreatment | 102 |
| | 6.4.2 Silane Treatment | 102 |
| | 6.4.3 Specimen Fabrication | 103 |
| 6.5 | Experimental Results for Kenaf Composites | 103 |
| | 6.5.1 Fourier Transform Infrared Spectroscopy Analysis | 103 |

        6.5.2    Mechanical Properties ........................................................................ 105
            6.5.2.1    Tensile Strength ................................................................ 105
            6.5.2.2    Flexural Strength............................................................... 106
        6.5.3    Scanning Electron Microscopy ....................................................... 106
6.6    Conclusions................................................................................................... 108
References............................................................................................................... 108

## 6.1    INTRODUCTION

Natural fibers can be extracted from plants, minerals, and animals. John and Thomas (2008) stated that the plant fibers may consist mostly of cellulose structures; meanwhile, animal fibers mostly contain protein structures. Aji et al. (2014), Anuar et al. (2011), and Mohanty et al. (2005) described that natural fibers have a low density property, are cost effective, have acceptable specific strength, have low carbon sequestration, and are highly biodegradable. Natural fibers are well-known due to their properties of non-abrasiveness during the treatment process, are easily amended to their surface, and are considered renewable (Aji et al. 2014; Chowdhury et al. 2013). Due to the advantages, fibers were also known to have superior specific strength and stiffness (Chowdhury et al. 2013). Thus, this attracted scientists and engineers for further research on natural fibers. This helps in increasing the use of natural fibers in composites due to their listed advantages (John and Thomas 2008).

The disadvantageous of natural fibers are their hydrophilic nature. Due to their hydrophilic nature, they have poor wettability with non-polar polymers (Chowdhury et al. 2013). However, this drawback can be overcome by different methods, such as the way to fabricate the fiber, restrain the choices of pairing the fiber and polymer, and chemical treatments that increase the fiber-matrix interfacial bonding (Chowdhury et al. 2013). According to Mohanty et al. (2005), natural fibers are categorized as cellulosic in nature. The major structures in the fibers are pectin, cellulose, hemicellulose, waxes, lignin, and pectin. Cellulose structure is made up of hydrophilic glucan polymers which consist of many 1,4-β-anhydroglucose units (Mohanty et al. 2005). Some polymer contains alcoholic hydroxyl groups which form intramolecular and intermolecular hydrogen bond structures with the macromolecular bond structure itself and also with other cellulose macromolecules or polar molecules. Thus, this indicates that the natural fibers were hydrophilic in nature.

Hemicellulose is different than cellulose and had several different sugar units compared to cellulose. Furthermore, hemicellulose had a pendant side groups in its branch chain and lower degree of polymerization than cellulose (John and Thomas 2008). Nevertheless, it still behaves as hydrophilic, easily hydrolyzed with acid and soluble in alkaline medium. The molecule hemicellulose tends to hydrogen bond with cellulose and plays a vital role in the structural component of the fiber cell (John and Thomas 2008). Lignin is described as biochemical polymer that acts as structural support in the plants which are made up of high molecular weight phenolic compounds, which are unaffected by microbial degradations (Mohanty et al. 2005; John and Thomas 2008). Lignin contains a complex, three-dimensional copolymer of aliphatic and aromatic constituents with a high molecular weight (John and Thomas 2007).

This component is insoluble in most solvents and cannot be fragmented into monomeric units. Compared to other structures, lignin is hydrophobic and amorphous in nature. It is known that there are few functional groups that may be present in lignin such as hydroxyl, carbonyl, and methoxyl groups. Based on John and Thomas (2007), there are five hydroxyl and five methoxyl groups per building unit found in lignin. The structures are sometimes assumed as derived from 4-hydroxy-3-methoxy phenyl-propane.

The present of pectin in plants makes the fibers flexible and usually consist of heteropolysaccaride collections. Waxes are formed due to different types of alcohol compounds. It is known that the natural fiber improved the impact as well as the deformation capacities that lead to the toughness of the composite (Udoeyo and Adetifa 2012). Fibers act as filler materials to improve the strength and toughness of the structure. The polymer functions as the bonding agent for the fibers to hold in place making the suitable structural components.

## 6.2 NATURAL FIBER COMPOSITES

### 6.2.1 Overview of the Composites

A composite material is defined as the combination of materials which are usually not found naturally (Mohanty et al. 2005). These materials are usually used in certain applications as they have been fabricated to reduce the energy consumption. Fiber matrix composites have some advantages, such as good corrosion resistance, a light weight, low thermal expansion, and high stiffness and strength when compared with metals such as aluminum and steel.

The main constraint in manufacturing composites is to find proper method in combining the matrix and reinforcement. A suitable method will produce better properties of the composites. There are several deficiencies that exist in the manufactured composites such as broken fibers, the unsuitable state of resin cure, and the presence of pores or voids in the matrix-rich regions (Harris 1999). One of the methods in manufacturing polymer composites is by molding process. According to Harris (1999), there are two major possible defects that are present in molding process. The first defect happens during controlling the resin curing time and flow. Due to this failure, the mold might not complete and poor adhesion could occur between fiber and polymer (Harris 1999). The second defect occurs in the composites that reduce the strength of the material due to the non-uniform filling or less attention to the flow patterns (Harris 1999).

Based on Begum and Islam (2013), the mechanical properties of a composite depend on several factors. Those factors are the nature of the resin, cross-linking agents, fiber treatment, resin-fiber adhesion, and also the processing technique of the composites. The raw fibers were added with liquid resin and the cross-linked catalysts that are used for hardening.

### 6.2.2 Fiber Composites

According to Faruk et al. (2012), the natural fibers in polymeric composites increased the various properties by fiber treatment. Examples of popular natural fibers used are bamboo, hemp, jute, sisal, kenaf, etc. In certain composite applications, natural fibers were found as an alternative substitute the glass fibers (Mohanty et al. 2005). Natural fibers had their own advantages in certain properties that make them superior and better than other materials that are usually used in the reinforced composites. However, there are some disadvantages of natural fibers that need to be considered and solved in order to have a better performance.

### 6.2.3 Benefits of Natural Fiber-Reinforced Composites

The major benefits of composite materials are due to its lightweight and durability properties. Other than that, composite materials are also classified as free from health hazards (Salleh et al. 2012). There are several advantages of using natural fibers in the reinforced composites compared to synthetic fibers, such as the superior acoustic insulation properties, and the renewable sources of the fibers. Furthermore, natural fibers are known to be safer and more beneficial to the environment. A proper mixture of matrix and reinforcement material can lead the proper desires for particular requirement in applications. Besides that, composites can be flexibly designed, as some of the composites are able to form complex shapes (Chollakup et al. 2012). According to Chollakup et al. (2012), the expenses of raw materials are not equivalent with the resulting products. The existence of natural fiber-reinforced composites allowed the industry to explore their hidden potential. An increase in the applications of these beneficial composites was seen as an alternative to replace carbon and glass fiber-reinforced materials. The increasing of interest in natural fibers were due to the shortage of unrenewable sources and the growing mineral source-based construction materials (Amran et al. 2014).

### 6.2.4 Drawbacks of Natural Fiber-Reinforced Composites

The main idea of composites is to replace plastic, glass, or other traditional reinforcement materials in the composites due to unbearable properties, such as in load-bearing applications. This was due to the lack of certain properties such as strength, stiffness, and dimensional stability. Recently, research using non-wood fibrous fiber such as kenaf, hemp, sisal, and flax commercially in combination with polypropylene in automotive, construction, and other industries. Begum and Islam (2013) had emphasized that the major drawback was due to the high moisture absorption of the natural fibers. This factor lead to an increase in the fragility of interfacial bonding that causes instability to its dimension and restrains the use of natural fibers. There are several drawbacks of natural fiber-reinforced composites discovered by John and Thomas (2007). One of them is the processing temperature for a composite which is limited to 200°C. This is because vegetation will start to degrade as they reach high temperature. Azwa and Yousif (2013) mentioned that the temperature had a greater influence to the performance of the natural fibers in the composites.

Another disadvantage of having natural fibers in the composite is that they had small microbial resistance and susceptibility to decay. This lead to the limitation of utilizing biofibers in its applications. Furthermore, natural fiber-reinforced composites had low durability especially on the natural fibers. Poor fire resistance and poor moisture resistance were also included in the consideration of using natural fibers in the composites. Other than that, natural fiber-reinforced composites had unknown environmental degradation towards the natural fiber-reinforced composites. Other factors that may affect the composites are ultraviolet radiation (UV), temperature, and humidity. Other than that, different natural fibers have different qualities and behaviors. Thus, the preparation methods of the raw fibers will be time consuming, so as to choose correct method and chemical substance for utilization.

### 6.2.5 Trends of Fiber Reinforced

According to John and Thomas (2007), they found that fiber-reinforced polymers were discovered in 1908. They stated in their article that productions of composites begun many thousand years ago. An example to this statement is the manufacturing of mud brick. Some of the fibers used in the fiber-reinforced polymer composites are fully synthetic fibers, such as glass, carbon, and aramid (Azwa and Yousif 2013; Begum and Islam 2013). After some time, problems such as high costs and waste disposal issues were arising. Another problem that has been issued by Azwa and Yousif (2013) is the emerging concept of green building. Due to these problems, they started to explore new material to replace the composite polymer. Hence, a number of researchers found that the biofiber-reinforced composites can be a replacement for the polymer composites (John and Thomas 2007). The types of composites depend on the processing and manufacturing processes. Thus, this showed that composites can be divided into three classes, which are green composites, hybrid composites, and textile composites (John and Thomas 2007).

Salleh et al. (2012) classified natural fiber-reinforced composites as lightweight and nonhazardous. Kenaf fibers were used in composites as health problems were on the rise because of the use of asbestos in reinforcement of cement and concrete matrices (Udoeyo and Adetifa 2012). According to Thiruchitrambalam et al. (2012) the replacement of synthetic fibers by natural fibers was due to environmental problems and health hazards during removal and manufacturing. Thus, some researchers started to find an alternative to overcome the problem. According to Cao et al. (2007), they identified that heat and alkali treatment to natural fibers will upgrade the mechanical properties of natural fibers.

### 6.2.6 Biocomposites

Biocomposites are composite materials that are made from the combination of natural fibers and petroleum-derived, nonbiodegradable polymers, or biopolymers (Mohanty et al. 2005; John and Thomas, 2007). The derivation of biocomposite is from biopolymer. Meanwhile, glass or carbon is from synthetic fibers. Other than that, biocomposites can also be categorized as a green and eco-friendly composite. Some researchers found that biocomposites can be a new solution for the environmental problems and also as a replacement for petroleum products. These composites looked as a potential and value-added source of income to the agricultural community (Mohanty et al. 2005). A combination of two or more biofibers with polymer matrix gives hybrid biocomposites. The efficiency of the fiber-reinforced composites also depends on the manufacturing process and the ability to transfer stress from the matrix and/or to the fiber (Mohd Yuhazri et al. 2011). Examples of natural fiber-reinforced composites are packages, trays, arm rests, instrument panels, doors, glove boxes, and seat backs.

### 6.2.7 Hybrid Composites

Hybrid composites are usually related to fiber-reinforced and resin-based materials. This type of composite is a mixture of two or more materials which explore their qualities as well as reducing their less required properties. Engineers are able to improve the original properties in these hybrid composites. The idea of using hybrid composites is that they can give many benefits to the users. One of them is that the composite can help to lessen the cost of composites containing reinforcements (Harris 1999).

### 6.2.8 Advantages of Natural Fibers in Composites

According to Begum and Islam (2013), as the sources of the material can easily and naturally be obtained, the energy needed to process the natural fibers is not as large as processing the synthetic fibers. Moreover, it cost-effective and lightweight as well. Because of the advantages of natural fibers, it is possible to substitute a large part of the synthetic fibers in an application. Another advantage of natural fibers was that it is easy to modify the surface of the natural fibers. The surface modification will increase the characteristic properties of the natural fibers and promote interfacial bonding.

### 6.2.9 Applications of Natural Fibers

Others products made from natural fibers like kenaf are ropes, twine, etc. Some of these natural fibers have been explored and used in applications. The reason behind this was due to the new properties of natural fibers, such as more comfortable fabrics. Another application of natural fibers can be seen in the automotive and aerospace sector. According to Anuar et al. (2011), natural fiber composites or biocomposites can help save fuel due to their lightweight properties.

## 6.3 FIBERS

### 6.3.1 Kenaf

*Hibiscus cannabinus* L. or kenaf is part of the hibiscus family. Kenaf plants have two types of fibers which is the outer bark or bast, which covers 40% of the plant, and an inner woody core which covers another 60% of the plant (Bazen et al. 2007). Nowadays, technologies are more advanced, and are able to separate the kenaf core and bark. According to Rashdi et al. (2010), previously kenaf fibers were used broadly as a jute-like material, which originates from the bast of the plant. This fiber can be found in most country such as Malaysia, India, Bangladesh, and Southeast Europe. In Malaysia itself, kenaf plants were seen to replace tobacco plants in the future (Meon et al. 2012).

Bahtoee et al. (2012) stated that the kenaf fiber is sensitive to photoperiods. Besides that, there are three different groups of cultivars which are early cultivar, mediate cultivar, and late maturity cultivar (Bahtoee et al. 2012). Those groups are distinct in terms of performance from each other. Besides that, kenaf is found to have better potential in reinforced fiber in thermosets and thermoplastics composites (Mohd Yuhazri et al. 2011). According to Ghani et al. (2012), kenaf fibers required less water to grow because it has an average yield of 1700 kg/ha with growing cycle of 150–180 days. Thus, this proves that every component of kenaf is useful. Kenaf were seen as high potential to reduce the environmental burden due to limited landfill facilities in Northern Nigeria (Udoeyo and Adetifa 2012). Kenaf were believed as one of the building materials.

### 6.3.2 Properties of Kenaf Fibers

Raw kenaf fiber is a bundle of lignocellulosic fibers which were obtained from the outer bark. The size of fiber bundle is influenced by the number of ultimate cells where lignin existed in between it. According to research done by Rashdi et al. (2010), it is found that kenaf was covered by 21.6% of lignin and pectin, 0.7% of cellulose, and the remaining by other components. Hence, to separate the fibers, lignin must be extracted from the fibers. The study carried out by Thiruchitrambalam et al. (2012) to prove that the cellulose will affect the strength of the composite.

Many studies have been carried to investigate the mechanical properties of kenaf fibers and its composites. According to the literature, kenaf has been declared as suitable replacement for reinforcement in biodegradable polymer composites (Thiruchitrambalam et al. 2012). Another study confirms that increasing the fiber content will cause the flexural properties of its composites to increase. (Thiruchitrambalam et al. 2012). For tensile properties, it shows that alkalization will increase the properties of the fiber. According to Moses et al. (2015), kenaf fibers were said to have excellent flexural and tensile strength compare to other natural fibers.

### 6.3.3 Advantages of Kenaf

According to Robinson (1988), kenaf was used in production of bags, rugs, rope, and twine. Kenaf fibers have their own benefits and weaknesses. As written by Bonnia et al. (2012), and Mohd Yuhazri et al. (2011), these fibers are biodegradable and ecologically friendly. Tahir et al. (2011) described kenaf as a white and strong fiber. Other than that, Tahir et al. (2011) has enhanced properties of kenaf fiber compare to wood fibers. It is found that kenaf fibers have a greater reinforcing effect on natural rubber compared to synthetic polyester fibers. Other than that, it is found that kenaf fibers have become a key factor in fabrication of composite and industrial applications (Mohd Yuhazri et al. 2011). Besides that, kenaf fibers are classified as a recyclable material and less expensive to produce than glass fibers. According to El-Shekeil et al. (2012), kenaf plants have good mechanical properties, short growing period, and high biomass output.

According to Hall et al. (1998), it is found that on the surface of the cellulose chains there are a huge number of hydroxyl groups present. With the presence of these groups, there is an improvement in the reaction and adherence between the epoxy resin/fiber interfaces. Nordin et al. (2013) verified the statements, where it is easy to do chemical or mechanical modification towards natural fibers, such as kenaf. Hence, the reinforcement using kenaf fiber is found to have potential in substituting glass fiber composites (Thiruchitrambalam et al., 2012). According to studied carried out by Zampaloni et al. (2007) as mentioned by Thiruchitrambalam et al. (2012), kenaf is used as reinforcement in maleated polypropylene, and shows that the composite has higher specific modulus than other natural fiber and E-glass.

### 6.3.4 Trend in Kenaf Composites

Previous researches were done by reinforcing the kenaf fibers with polylactic acid (PLA), polypropylene (PP), and ethylene-propylene-diene monomer (EPDM), nylon and other inexpensive polymers (Anuar et al. 2011; Jeyanthi and Rani 2012). According to Jeyanthi and Rani (2012), kenaf has proved to have better mechanical and thermal properties when mix with PP. Huda et al. (2008) has stated that coupling agent was used to increase the degree of cross-linking in the interface section. For example, Huda et al. (2007), carried out using 3-aminopropyltriethoxysilane (APS) as the coupling agent since APS has ability to form bond to polylactide. APS consists of ethoxy groups which hydrolyzed in water and able to form stable covalent bonds to the cell wall.

Another research was done for the composite of kenaf fibers with PLA as reported by Serizawa et al. (2005), and Huda et al. (2008). However, the composite was only tested on building material and automobile. At the end of the research, combination of kenaf with PLA has improved some of the properties such as heat resistance and modulus of PLA (Serizawa et al. 2005). This proves kenaf has similar synergistic effects as inorganic fillers. Amel et al. (2014) carried out studies of the effects of kenaf bast fibers on hydration behavior of cement. The hydration manners were explored through its characteristics of highest hydration temperature and the time needed to reach that temperature. Both authors found that high inhibition materials such as NaOH and benzoate-treated fibers have bigger influence towards the fibers. This concludes that both materials are not suitable for fiber extraction as compared to intermediate inhibition such as $CaCl_2$, $AlCl_3$, and CaO.

Tahir et al. (2011) has mentioned that a study about liquefied kenaf core (LKC) was used as a polyol in synthesize polyurethane adhesive. The adhesive produced showed good performance which is seen as a potential for edge-gluing. In a study by Meon et al. (2012), they have treated kenaf fibers with sodium hydroxide. According to Mohd Yuhazri et al. (2011), efficiency of the fiber-reinforced composites is influenced on the manufacturing procedure. In other words, it was the process of transferring stress from the polymer matrix to the fibers. A study carried by Mohd Yuhazri et al. (2011) used polyester resin in reinforcing the kenaf fibers, which underwent a vacuum infusion method. The study was to examine the mechanical properties. Accoridng to Aji et al. (2014), using sodium hydroxide (NaOH) in the modification of kenaf fibers with different concentrations improved their mechanical properties. Some of the researchers thought that chemical treatment on the natural fibers would clean the surface of the fibers and break the moisture absorption process. Aji et al. (2014) also found that the roughness of the natural fibers' surface will increase with the alkaline optimal concentration of 6%. Aji et al. (2014) carried out a study of combining kenaf with pineapple leaf fibers (PALF). This combination had high mechanical properties as well as raised loading in the composite. The flexural and torsional rigidity in PALF was almost the same with jute fibers (Aji et al. 2014).

Another study of these natural fibers was carried out by Diharjo et al. (2013). They studied kenaf fiber-reinforced PP composites with and without clay. The manipulated variable in this study was the composition of PP/kenaf and PP/kenaf/clay. The clay used in the study showed that the present of clay in the composites prevented the development of PP fiber bonding to occur. Some of the studies used kenaf fibers as reinforcement on the properties of concrete. They mixed up about 0.25% and 0.40% of kenaf fibers in foamed concrete. Then, the composite concrete was tested for flexural strength, water absorption, and compressive strength. The results showed that the properties increased about 61% and 78% for both 0.25% and 0.40%. They also included an increase of 7% and 9%, respectively, for the same kenaf fiber concrete. However, the compressive strength of kenaf fiber concrete had slightly decreased (Moses et al. 2015). There was a study about comparing tensile properties of bamboo, jute, and kenaf done by Hojo et al. (2014). According to Hojo et al. (2014), all the natural fibers used unsaturated polyester (UP) as a matrix, which were bamboo/UP, jute/UP, and kenaf/UP. The composites were prepared by hand lay-up and compression molding techniques. The results from the research found that kenaf/UP has better tensile modulus and ultimate strength compared to bamboo/UP and jute/UP. Two more types of composite have similar tensile modulus and ultimate strength.

Ali et al. (2015) carried out research on the effect of single- and double-stage chemically-treated kenaf fibers. In this study, kenaf fibers underwent chemical treatment separately in single and double stages. The first stage was that the kenaf fibers were treated with $Cr_2(SO_4)_3.12(H_2O)$, whereas in the double stage, kenaf fibers were treated with $CrSO_4$ and $NaHCO_3$. The purpose of both treatments is to enhance the compatibility and the adhesion between the kenaf fiber and PVA matrix. The composites with the treated kenaf fibers had a higher tensile strength than composites with untreated kenaf fibers. At the end, it is found that the double-stage treatment gives better mechanical properties and lower moisture intake than the single stage treatment. An examination was conducted by Islam et al. (2010) using laccase in modification of kenaf fibers surface. The raw and treated kenaf fibers used recycled polypropylene as the matrix. The treatment parameters are enzyme concentration and soaking time. The highest tensile properties were obtained from the optimum fiber loading of 40%. Other properties such as flexural, impact, and thermal increase due to the treatment were carried out. Moisture absorption was reduced by 37% with the aid of coupling agent and morphology investigation shows better adhesion in the composites.

Combinations of steam–chemical–ultrasonic treatment and chemical–ultrasonic treatment were used to treat kenaf fibers. The results showed that a higher degree of crystallinity and better fibrillation were produced by the fibers treated with steam. The fiber surface roughness was reduced under steam pretreatment. The fibers treated without steam show there is a formation of nanofibers on the surface. The purpose of steam treatment is to stimulate cleavage of the noncellulosic components and enhance the elimination of the noncellulosic components and the fibrillation process.

### 6.3.5 The Drawback of Kenaf to Its Composites

There is many disadvantages of kenaf fibers that must be considered, which are found in other natural fibers. According to Anuar et al. (2011), poor compatibility between natural fibers and matrices happened when natural fibers were used in composites. This fact can be supported by Westman et al. (2010) which stated that the recent issues with natural fibers composites were resin compatibility and water absorption. Problems have arisen in the processing of kenaf due to the roughness and toughness of the fiber bundle. Besides that, natural fibers such as kenaf itself have a low thermal resistance (Azwa and Yousif 2013). Hence, to have a composite with good mechanical properties, kenaf and other natural fibers must be treated by suitable chemical treatments. Mohd Yuhazri et al. (2011) also emphasizes implementing fiber adjustment for better mechanical properties.

## 6.4 CHEMICAL TREATMENT

Treatment procedures were planned and grouped based on the main processes of pretreatment of kenaf fibers, silane treatments, and the specimen's fabrication. The raw material was separated into two parts: untreated and treated raw material. There are two chemical treatment involved in the procedure; pretreatment and silane treatment.

### 6.4.1 Pretreatment

Ethanol (900 mL) and water (100 mL) was poured in a basin, with a ratio of ethanol–water that was 9:1. The solution was stirred until it well mixed. Then, the raw kenaf fibers were put into the basin, and stirred. The solution was then prepared again with same ratio to ensure all the raw material completely immersed in the solution. The raw material was allowed to immerse in the solution for 12 hrs. After that, the treated material was being filtered in a conical flask using a filter paper. There are four sets of conical flask. After filtering process, the material underwent rinsing process with at least 4 L of distilled water. Then, the material was left for drying in the oven set at temperature of 65°C for 12 hrs.

### 6.4.2 Silane Treatment

The treatment used chemical compound which contains silane ($SiH_4$) as coupling agents. Another solution with ratio of ethanol–water of 9:1 was being prepared in a basin. The dry material was put into the solution with fully immersed. About 2 mL of triethoxysilane with concentration of 95% was poured into the solution. The solution was vigorously shaken. The pH of the solution was found to be 6, which was tested using pH paper. About 20 mL of acetic acid with concentration of 100% was poured into the solution to achieve the pH of 4. Then, the solution was well stirred, and tested for its pH. The solution with the material was divided into three equal amounts into a conical flask. All conical flasks were then placed into a chemical laboratory shaker. The shaker was set at operational speed at 80 rpm for 24 hrs, its running temperature at 26.7°C and holding temperature at 28°C for the operation. Then, the material was poured into the filter funnels for undergo filtering and rinsed again. The material was put in the oven for drying purposes.

### 6.4.3 Specimen Fabrication

A combination of hand lay-up and compression molding techniques were used to prepare the kenaf fiber-reinforced unsaturated polyester biocomposites. First, the mold was washed and cleaned with ethanol, rinsed with water, and dried in the oven. Then, the mold was sprayed with multipurpose silicone spray for easy removal. 10 gm of dry material was added to 190 gm of liquid unsaturated polyester resin to form 200 gm of mixture. Then, 4 gm of hardener was added to the mixture. The mixture was well mixed, and then poured into the mold. The mixture was spread to ensure the mixture fill the entire mold. The mold was compressed by hand before being tightened by the screws at the four curves with nuts. The mixture was left for 2 hrs during the drying process. The solidified sample was taken out from the mold by using a putty knife and rubber hammer. The sample was cropped from thin film by using scissors. The above-mentioned procedures are repeated with different ratios of dry material and liquid unsaturated polyester resin to produce samples with 15%, 20%, and 25% wt dry material, respectively.

## 6.5 EXPERIMENTAL RESULTS FOR KENAF COMPOSITES

### 6.5.1 Fourier Transform Infrared Spectroscopy Analysis

The IR spectrum of raw kenaf fibers and treated kenaf fibers are shown in Figures 6.1 and 6.2. For all the first peaks in different weight of both raw and treated kenaf fibers, the peaks were observed at 2962.66 cm$^{-1}$. This value is corresponds to the C-H group stretching vibration. For the peaks that observed at 1718.58 cm$^{-1}$ and 1720.50 cm$^{-1}$, they satisfied the regions of the carbonyl group stretching (C=O bond). The regions for carbonyl group are between 2000 and 1000 cm$^{-1}$ (Coates 2000; Ali et al. 2015). Meanwhile, strong C=O bond absorption that represents aldehydes and ketones is in regions from 1660 to 1770 cm$^{-1}$. With these bands shows in the spectroscopy, it confirmed that lignin structure is present in the kenaf fibers (Ali et al. 2015). Starting from second peak and onwards, different frequencies are shown in the spectroscopy for different weight of both raw and treated kenaf fibers. Compared with raw and treated of 5 g kenaf fibers, the second peak of raw kenaf fibers is 2916.37 cm$^{-1}$ and second peak of treated kenaf fibers is 1718.58 cm$^{-1}$. It shows there is significantly decreased between both raw and treated kenaf fibers. The same situation occurs at third peak of 10 g treated kenaf fibers, second peak of 15 g treated kenaf fibers and second peak of 20 g treated kenaf fibers. This indicates that the degradation or modification of lignin and hemicellulose structures (Abdul Razak et al. 2014).

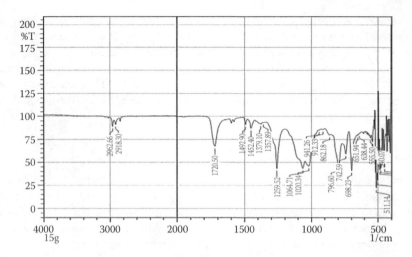

**FIGURE 6.1** IR spectrum of composites of raw kenaf fiber.

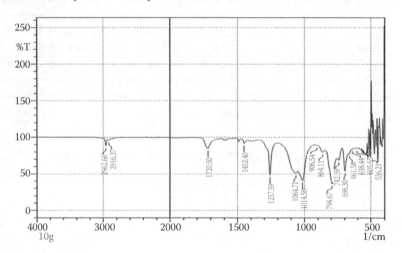

**FIGURE 6.2** IR spectrum of composites of treated kenaf fiber.

Besides that, the changes also indicate that the elimination of wax and pectin has occurred (Abdul Razak et al. 2014).

According to Coates (2000), the range for methyl C-H is 2970–2950 cm$^{-1}$ for asymmetry stretch or 2880–2860 cm$^{-1}$ for symmetry stretch. A weaker methyl or methylene band shows chain branching, and strong methyl bands indicates significant splitting (Coates 2000). As mentioned, there is carbonyl group in the composites. The carbonyl groups can be ketones, aldehydes, or esters. The carbonyl groups contain the C=O functional group. If ketones exist, the bond of C-H stretching is broader than the stretching of C-H bonds in aldehydes. From the results, the region shows the peaks at 1718.58 cm$^{-1}$ and 1720.50 cm$^{-1}$, which demonstrates that the ketones are present. However, according to Coates (2000), conjugations play as an important role when observing frequency of carbonyl. Which means the conjugations will determine that they either connect to aromatic rings or conjugate to C=C or C=O. Besides that, the level of carbonyl frequency also will help to differentiate the types of carbonyl compounds, in which it will determine if the carbonyl group will directly or indirectly attach to aromatic rings (Coates 2000).

### 6.5.2 Mechanical Properties

#### 6.5.2.1 Tensile Strength

Tensile strength values of raw kenaf fiber UP composites (RK-UP-C) and treated kenaf UP Composites (TK-UP-C) at different fiber loading are shown in Figure 6.3. For RK-UP-C, the tensile strength increased with an increase in fiber loading up to 15% and decreased with further increase the fiber loading (Rahman et al. 2008, 2009; Islam et al. 2010). A similar trend was observed for TK-UP-C. As the fiber load increased, the weak interfacial area between the fiber and matrix increased, which in turn decreased the tensile strength.

In order to increase mechanical properties of the composites, the kenaf fibers were treated by silane. Due to the elimination of hydroxyl group from the kenaf fibers, the interfacial bonding between the fiber and the UP matrix increased in the resultant composites. The tensile strength of the 15% treated fiber loaded composites were higher followed by raw fiber-reinforced composites, respectively (Figure 6.3). The tensile strength of the treated fiber-loaded composites then decreased with an increase in fiber loading due to the increase in weak interfacial area, as mentioned above.

Figure 6.4 shows the deviation of the Young's modulus at different fiber loading. The Young's modulus increased with an increase in fiber loading (Paul et al. 2003; Rahman et al. 2008, 2009). During tensile loading, moderately disconnected microspaces are created, which hampers stress propagation between the fiber and the UP matrix (Rahman et al. 2008, 2009; Paul et al. 2003). As the fiber loading increases, the element of barrier increases, which in turns increases the toughness. Again, there was a high increase in toughness in treated fiber-reinforced composites when compared with raw fiber-reinforced composites, respectively.

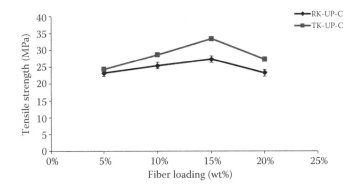

**FIGURE 6.3** Tensile strength of RK-UP-C and TK-UP-C at different fiber loading.

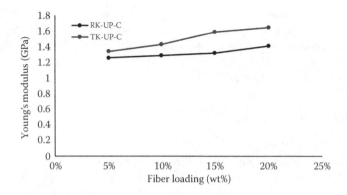

**FIGURE 6.4** Young's modulus of RK-UP-C and TK-UP-C at different fiber loading.

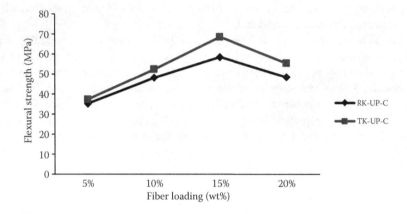

**FIGURE 6.5** Flexural strength of RK-UP-C and TK-UP-C at different fiber loading.

### 6.5.2.2 Flexural Strength

Flexural strength of raw and treated kenaf fiber-reinforced UP composites at different filler loading are shown in Figure 6.5. The flexural strength increased with an increase in fiber loading up to 15%, however for 20% fiber loading, the flexural strength decreased, which was shown in Figure 6.5. Treated kenaf fiber composites yielded higher flexural strength values compared with raw fiber-reinforced UP composites. Similar results were also reported by other researchers (Lee et al. 2004). The flexural strength results proved that the treatment enhanced the interfacial bonding between fiber and matrices (Wong et al. 2010).

### 6.5.3 Scanning Electron Microscopy

The scanning electron microscopy (SEM) micrographs of RK-UP-C and TK-UP-C fracture surfaces are shown in Figures 6.6 and 6.7, respectively.

From the Figure 6.6, SEM micrograph of the RK-UP-C was detected to have a non-uniform surface with less pores. Besides, there are small cracks between the fiber and the

Impact of Silane Treatment on the Properties of Kenaf Fiber

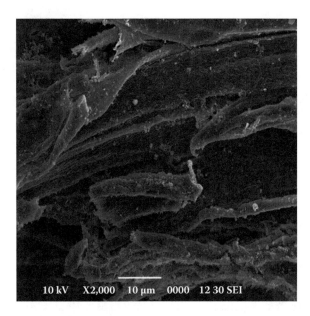

**FIGURE 6.6** SEM micrographs of RK-UP-C.

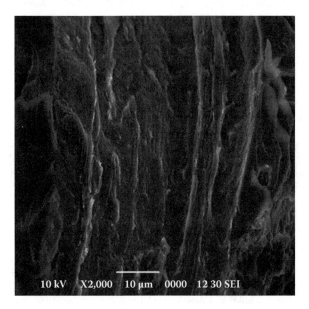

**FIGURE 6.7** SEM micrographs of TK-UP-C.

matrix. This indicates that the polymer-fiber interaction was not strong and thus the filler had less compatibility with the UP matrix. On the other hand, the TK-UP-C showed better adhesion between fiber and matrix, which was shown in Figure 6.7 (Zhang et al. 2013b). This is due to the strong interfacial bonding between fiber and matrix that allowed effective stress transfer from the matrix to the fiber in composite (Wong et al. 2010).

## 6.6 CONCLUSIONS

The composites of raw and treated kenaf fibers were produced with unsaturated polyester resin as the polymer matrix. In Fourier Transform Infrared spectroscopy (FTIR), the peaks in the treated kenaf fiber were reduced the hydroxyl group (OH) and remove the wax and lignin. A better tensile strength and tensile modulus as well as flexural strength was performed by both the raw and treated kenaf fiber composites loading up to 15% wt. The treated kenaf fiber composites performed better than raw kenaf fibers, as the silane treatment eliminated the hydroxyl groups from the kenaf fibers, which had enhanced the interfacial bonding between the fiber and UP matrix. The strong interfacial bonding between the fiber and matrix improved the surface morphology of the treated kenaf fibers.

## REFERENCES

Abdul Razak, N.I., Ibrahim, N.A., Zainudin, N., Rayung, M., and W.Z. Saad. 2014. The influence of chemical surface modification of kenaf fiber using hydrogen peroxide on the mechanical properties of biodegradable kenaf fiber/poly (lactic acid) composites. *Molecules* 19, no. 3: 2957–2968.

Aji, I.S., Zainudin, E.S., Sapuan, S.M., Khalina, A., and M.Z. Khairul. 2014. Effect of fibre/matrix modification on tensile properties and water absorption behavior of hybdiridzed kenaf/PALF reinforced HDPE composite. *Research & Reviews: Journal of Engineering and Technology* 3, no. 3: 1–8.

Ali, M.E., Yong, C.K., Ching, Y.C., Chuah, C.K., and N.S. Liou. 2015. Effect of single and double stage chemically treated kenaf fibers on mechanical properties of polyvinyl alcohol film. *BioResources* 10, no. 1: 822–838.

Amel, B.A., Paridah, M.T., Rahim, S., Osman, Z., Zakiah, A., and S.H. Ahmed. 2014. Effects of kenaf bast fibres on hydration behaviour of cement. *Journal of Tropical Forest Science* 26, no. 3: 340–346.

Amran, M., Izamshah, R., Hadzley, M., Shahir, M., Amri, M., Sanusi, M., and H. Hilmi. 2014. The effect of binder on mechanical properties of kenaf fibre/polypropylene composites using full factorial method. *Applied Mechanics and Materials* 695, no. 1: 709–712.

Anuar, H., Hassan, N.A., and F. Mohd Fauzey. 2011. Compatibilized PP/EPDM-Kenaf fibre composite using melt blending method. *Advanced Materials Research* 264–265, no. 1: 743–747.

Azwa, Z.N. and B.F. Yousif. 2013. Thermal degradation study of kenaf fibre/epoxy composites using thermos gravimetric analysis. Paper presented at 3rd Malaysian Postgraduate Conference (MPC2013), Malaysia.

Bahtoee, A., Zargari, K., and E. Baniani. 2012. An investigation on fiber production of different kenaf (*Hibiscus cannabinus* L.) genotyopes. *World Applied Sciences Journal* 16, no. 1: 63–66.

Bazen, E., Roberts, R.K., and B.C. English. 2007. Economic Feasibility of Kenaf Production in Three Tennessee Counties. Paper presented at Agricultural Economics Association Annual Meeting, Alabama.

Begum, K., and M.A. Islam. 2013. Natural fiber as substitute to synthetic fiber in polymer composites: A review. *Research Journal of Engineering Sciences* 2, no. 3: 46–53.

Benyahia, A., Merrouche, A., Rokbi, M., and Z. Kouadri. Study the effect of alkali treatment of natural fibers on the mechanical behavior of the composite unsaturated polyester polyester-fiber alfa. 21ème Congrès Français de Mécanique, Bordeaux.

Bergeret, A. 2011. Environmental-friendly biodegradable polymers and composites. *Integrated Waste Management –Volume 1*, Ed. S. Kumar. European Union: InTech.
Bonnia, N.N., Mahat, M.M., Surip, S.N., Anuar, H., Hassan, N.A., and S. Ahmad. 2012. Polyester/kenaf composite; effect of matrix modification. Paper presented at Business, Engineering, and Industrial Applications (ISBEAIA), 2012 IEEE Symposium, Indonesia.
Cao, Y., Sakamoto, S., and K. Goa. 2007. Effects of heat and alkaline treatments on mechanical properties of kenaf fibers. Paper presented at 16th International Conference on Composite Materials, Japan.
Chollakup, R., Smithhipong, W., and P. Suwanruji. 2012. Chapter 8: Environmentally friendly coupling agents for natural fiber composites. *Natural Polymers, Vol. 1: Composites*, Ed. M. J. John and S. Thomas. Cambridge: Royal Society of Chemistry.
Chowdhury, M.N.K., Beg, M.D.H., and M.R. Khan. 2013. Biodegradability of nanoparticle modified fiber reinforced polyester resin nanocomposite. *Procedia Engineering* 68, no. 1: 431–438.
Cicala, G., Cristaldi, G., Recca, G., and A. Latteri. 2010. Composites based on natural fiber fabrics. *Woven Fabric Engineering*, Ed. P. D. Dubrovski. European Union: InTech.
Coates, J. 2000. Interpretation of Infrared Spectra, A Practical Approach. *Encyclopedia of Analytical Chemistry*, Ed. R. A. Meyers. United States of America: John Wiley and Sons.
Davallo, M., Pasdar, H., and M. Mohseni. 2010. Mechanical properties of unsaturated polyester resin. *International Journal of ChemTech Research* 2, no. 4: 2113–2117.
Dholakiya, B. 2012. Unsaturated polyester resin for specialty applications. *Polyester*, Ed. H. E. D. Saleh. European Union: InTech.
Diharjo, K., Hastuti, S., Triyasmoko, A., Sumarsono, A.G., Putera, D.P., Riyadi, F., Probotianto, Y.C., and M. Nizam. 2013. The application of kenaf fiber reinforced polypropyelene composites with clay particles for the interior panel of electrical vehicle. Paper presented at 2013 Join International Conference on Rural Information & Communication Technology and Electric-Vehicle Technology (rICT & ICeV-T), Indonesia.
El-Shekeil, Y.A., Sapuan, S.M., Khalina, A., Zainudin, E.S., and O.M. Al-Shuja'a. 2012. Influence of chemical treatment on the tensile properties of kenaf fiber reinforced thermoplastic polyurethane composite. *EXPRESS Polymer Letters* 6, no. 12: 1032–1040.
Faruk, O., Bledzki, A.K., Fink, H.P., and M. Sain. 2012. Biocomposites reinforced with natural fibers: 2000–2010. *Progress in Polymer Science* 37, no. 11: 1552–1596.
Ghani, M.A.A., Salleh, Z., Hyie, K.M., Berhan, M.N., Taib, Y.M.D. and M.A.I. Bakri. 2012. Mechanical Properties of Kenaf/Fiberglass Polyester Hybrid Composite. *Procedia Engineering* 41, no. 1: 1654–1659.
Gon, D., Das, K., Paul, P., and S. Maity. Jute composites as wood substitute. *International Journal of Textile Science* 1, no. 6: 84–93.
Hall, H.L., Bhuta, M., and J.M. Zimmerman. 1998. Kenaf Fiber Reinforced Composite Athletic Wheelchair. Paper presented at 17th Southern Biomedical Engineering Conference, USA.
Hamma, A., Kaci, A., Mohd Ishak, Z.A., and A. Pegoretti. 2014. Starch-grafted-polypropylene/kenaf fibres composites. Part 1: Mechanical performances and viscoelastic behavior. *Composites Part A: Applied Science and Manufacturing* 56, no. 1: 328–335.
Harris, B. (1999). *Engineering Composite Materials*. London: The Institute of Materials.
Hojo, T., Xu, Z., Yang, Y., and H. Hamada. 2014. Tensile properties of bamboo, jute and kenaf mat-reinforced composite. *Energy Procedia* 56, no. 1: 72–79.
Hosier, I.L., Vaughan, A.S., Mitchell, G.R., Siripitayananon, J., and F.I. Davis. 2004. Polymer Characterization. In *Polymer Chemistry: A Practical Approach*. Ed. F.I. Davis. United Kingdom: Oxford University Press.

Huda, M.S., Drzal, L.T., Mohanty, A.K., and M. Misra. 2008. Effect of fiber surface-treatments on the properties of laminated biocomposites from poly(lactic acid) (PLA) and kenaf fibers. *Composites Science and Technology* 68, no. 2: 424–432.

Islam, M.N., Rahman, M.R., Haque, M.M., and M.M. Huque. 2010. Physico-mechanical properties of chemically treated coir reinforced polypropylene composites. *Composites Part A* 41, no. 2: 192–198.

Jeyanti, S. and J.J. Rani. 2012. Improving mechanical properties by kenaf natural long fiber reinforced composite for automotive structures. *Journal of Applied Science and Engineering* 15, no. 3: 275–280.

John, M.J. and S. Thomas. 2008. Biofibres and biocomposites. *Carbohydrate Polymers* 71, no. 3: 343–364.

Ku, H., Wang, H., Pattarachaiyakoop, N., and M. Trada. A review on the tensile properties of natural fiber reinforced polymer composites. *Composites Part B: Engineering* 42, no. 4: 856–873.

Lee, S.H., Kim, J.E., Song, H.H., and S.W. Kim. 2004. Thermal properties of maleated polyethylene/layered silicate nanocomposites. *International Journal of Thermophysics* 25: 1585–1595.

Meon, M.S., Othman, M.F., Husain, H., Remeli, M.F., and M.S. Mohd Syawal. 2012. Improving Tensile Properties of Kenaf Fibers Treated with Sodium Hydroxide. *Procedia Engineering* 41, no. 1: 1587–1592.

Mohanty, A.K., Misra, M., and L.T. Drzal. 2005. *Natural Fibers, Biopolymers and Biocomposites.* Boca Raton, FL: Taylor and Francis Group.

Mohd Yuhazri, Y., Phongsakorn, P.T., Sihombing, H., Jeefferie, A.R., Puvanasvaran, P.K.A.M., and K. Rassiah. 2011. Mechanical properties of kenaf/polyester composites. *International Journal of Engineering & Technology* 11, no. 1: 127–131.

Moses, O.T., Samson, D., and O.M. Waila. 2015. Compressive strength characteristics of kenaf fibre reinforced cement mortar. *Advances in Materials* 4, no. 1: 6–10.

Nordin, N.A., Md Yussof, F., Kasolang, S., Salleh, Z., and M.A. Ahmand. 2013. Wear rate of natural fibre: Long kenaf composite. *Procedia Engineering* 68, no. 1: 145–151.

Osman, E., Vakhquelt, A., Mutasher, S., and I. Sbarski. 2012. Effect of water absorption on tensile properties of kenaf fiber unsaturated polyester composites. *Suranaree Journal of Science Technology* 20, no. 3: 183–195.

Paul, W., Jan, I., and Ignaas, V. 2003. Natural fibers: Can they replace glass in fiber reinforced plastics. *Composites Science and Technology* 63, no. 2: 1259–1264.

Rahman, M.R., Huque, M.M., Islam, M.N., and M. Hasan. 2008. Improvement of physic-mechanical properties of jute fiber reinforced polypropylene composites by post-treatment. *Composites Part A* 39: 1739–1747.

Rahman, M.R., Huque, M.M., Islam, M.N., and M. Hasan. 2009. Mechanical properties of polypropylene composites reinforced with chemically treated abaca. *Composites Part A* 40, no. 4: 511–517.

Rashdi, A.A.A., Sapuan, S.M., Ahmad, M.M.H.M., and A. Khalina. 2010. Combined effects of water absorption due to water immersion soil buried and natural weather on mechanical properties of kenaf fibre unsaturated polyester composites (KFUPC). *International Journal of Mechanical and Materials Engineering (IJMME)* 5, no. 1: 11–17.

Robinson, F.E. 1988. Kenaf: A new fiber crop for paper production. *California Agriculture* 42: 31–32.

Salleh, Z., Taib, Y.M., Hyie, K.M., Mihat, M., Berhan, M.N., and M.A.A. Ghani. 2012. Fracture toughness investigation on long kenaf/woven glass hybrid composite due to water absorption effect. *Procedia Engineering* 41, no. 1: 1667–1673.

Serizawa, S., Inoue, K., and M. Iji. 2005. Kenaf fiber-reinforced biomass-plastics used for electronic products. Paper presented at Fourth International Symposium on Environmentally Conscious Design and Inverse Manufacturing, 2005 Eco Design, Japan.

Shivnand, H.K., Inamdar, P.S., and G. Sapthagiri. 2010. Evaluation of tensile and flexural properties of hemp and polypropylene based natural fiber composites. Paper presented at 2010 2nd International Conference on Chemical, Biological and Environmental Engineering (ICBEE), Egypt.

Tahir, P.M., Ahmed, A.B. Saiful Azry, S.O.A., and Z. Ahmed. 2011. Retting process of some bast plant fibres and its effect on fibre quality: A review. *Bio Resources* 6, no. 4: 5260–5281.

Thiruchitrambalam, M., Alavudeen, A., and N. Venkateshwaran. 2012. Review on kenaf fiber composites. *Reviews on Advanced Materials Science* 32, no. 2: 106–112.

Udoeyo, F.F., and A. Adetifa. 2012. Characteristics of kenaf fiber-reinforced mortar composites. *International Journal of Research & Reviews in Applied Sciences* 12, no. 1: 18–26.

Westman, M.P., Fifield, L.S., Simmons, K.L., Laddha, S.G., and T.A. Kafentzis. 2010. *Natural fiber composites: A Review*. Washington: Pacific Northwest National Laboratory.

Wong, K.J., Yousif, B.F., Low, K.O., Ng, Y., and S.L. Tan. 2010. Effects of fillers on the fracture behavior of particulate polyester composites. *Journal of Strain Analysis for Engineering Design* 45: 67–78.

Zhang, H.H., Cui, Y., and Z. Zhang. 2013. Chemical treatment of wood fiber and its reinforced unsaturated polyester composites. *Journal of Vinyl and Additive Technology* 19, no. 1: 18–24.

# 7 Effects of Material Types on the Failure Modes Crashworthiness Parameters of Kenaf Composite Hexagonal Tubes

*M.F.M. Alkbir, S.M. Sapuan,
A.A. Nuraini, and M.R. Ishak*

**CONTENTS**

| | |
|---|---|
| 7.1 Introduction | 114 |
| 7.2 Experimental Section | 115 |
|     7.2.1 Geometry and Mandrel Design | 115 |
|     7.2.2 Fabrication Process | 117 |
| 7.3 Crush Testing and Crashworthiness Parameters | 118 |
| 7.4 Results and Discussion | 118 |
|     7.4.1 Load-Displacement Curves and Failure Mechanisms | 118 |
|         7.4.1.1 Results of CFKM | 118 |
|         7.4.1.2 Results of KYKM | 119 |
|         7.4.1.3 Results of GFKY | 120 |
|         7.4.1.4 Results of CFKY | 121 |
|         7.4.1.5 Results of GFKM | 121 |
|     7.4.2 Results of Crashworthiness | 122 |
|         7.4.2.1 Effect of Hybrid Material in Initial Peak Load | 124 |
|         7.4.2.2 Catastrophic Failure Mode Indicator | 125 |
| 7.5 Conclusions | 126 |
| 7.6 Funding | 126 |
| References | 127 |

## 7.1 INTRODUCTION

Various investigations have shown that there are forms of the failure mode in composite materials, in both fiber and matrix combinations. The majority of this research has focused on axial compression loading. Crashworthiness parameters and the failure modes of fiber-reinforced composite tubes (FTR) are dependent on several parameters, for example, the matrix and fiber types, the specimen geometry, and the manufacturing methods.

Altering any of these parameters undoubtedly causes two changes in the specific energy absorption capacity of composite structures (Abdewi et al. 2008; Mamalis et al. 2004; Melo et al. 2008; Palanivelu et al. 2010; Shibata et al. 2006; Warrior et al. 2008; Hull et al. 1991; Meran et al. 2015). Carbon, kevlar, and carbon-kevlar hybrid fiber composites with epoxy resin have all been used to manufacture composite circular tubes. Results have shown that the plain weave carbon/epoxy type is the most appropriate for automobile body structures due to its high energy absorption specific energy absorption (SEA) Abosbaia et al. (2003, 2005) studied the effect of segmentation and non-segmentation on the crushing behavior, energy absorption, and failure mechanisms of composite tubes under an axial crushing load and a lateral crushing load. Three forms of fiber reinforcements were used—tissue mat, glass fibers, carbon fabric fibers, and cotton fabric fibers. Results demonstrated that a modification in the segmentation sequence influences the crushing load significantly, while the segmented composite tubes made from carbon fabric fibers and cotton fabric fibers exhibited improved smart energy absorption capabilities in addition to a stable load-carrying capability. Mahdi et al (2014) experimented with five types of hybrid circular–cylindrical composite shells, and the relationship of glass/carbon sequencing to the energy absorption capability. They showed that the glass–carbon–glass/epoxy (GCG) segmentation offers the greatest energy absorption when compared to glass–glass–carbon/epoxy (GGC) and carbon–glass–glass/epoxy (CGG). The reason for this was attributed to the failure of the glass fiber layers under the transverse and axial shear. The carbon layers also buckled, and the GGC– and the CGG–segmented composites failed when the glass fiber layers deboned and the carbon layer cracked.

Some investigation has been carried out into the use of natural fibers as an environmentally supportable alternative to conventional glass and carbon fiber composites, since fibers derived from plants are renewable. Natural fibers are also low-cost, have a low density and high specific properties, are non-abrasive, and are less harmful during handling (Mohanty et al. 2001; Alkbir et al. 2016; Saheb et al. 1999; Yan et al. 2014; El-Shekeil et al. 2014).

Dittenber and Gangarao (2012), compared 20 of the most frequently used natural fibers (e.g., sisal, ramie, kenaf, jute, hemp, flax, coir, cotton, etc.) with glass fibers in specific modulus, with particular focus on the cost per weight and the cost per unit length (which is capable of resisting a 100-kN load). They concluded that flax fibers offer the best potential combination of properties—low cost, light weight, and high strength and stiffness—for structural applications.

Natural fibers are generally used as the fiber-reinforcemed composite material in composites such as kenaf/epoxy composites (El-Shekeil et al. 2014; Yousif et al. 2012), sisal–jute fiber-reinforced polyester composites (Ramesh 2013), and woven natural silk/epoxy rectangular composite tubes (Eshkoor et al. 2013; Oshkovr et al. 2012; Ataollahi et al. 2012). Studies have recommended that natural fibers have great potential to be used in composite materials along with synthetic fibers. Meredith et al. (2012) used three types of natural fiber—unwoven hemp, woven flax, and woven jute—to examine the issue of the natural fiber reinforcement of composite energy absorbing structures. The results showed that unwoven hemp exhibited the highest value of SEA (54.3 J/g) while woven flax had an SEA of 48.5 J/g and woven jute had an SEA of 32.6 J/g. Kenaf (*Hibiscus cannabinus* L.) has become an important commodity in Malaysia and is widely utilised in the manufacturing, packaging, automotive, and furniture industries. Kenaf fibers are recognized as being some of the finest natural fibers for automotive applications, because the source plant produces fibers four times as quickly when compared to other sources like jute and wood (Shibata et al. 2006; Sapuan et al. 2013; Maepa et al. 2015). Chapter 8 illustrates the effect of the fiber-reinforcement type on the failure modes of composite hexagonal tubes.

## 7.2 EXPERIMENTAL SECTION

The experimental work has been divided into three sections as follows: The first section involves a mandrel design and fabrication, which deals with the hexagonal tube shape. The second section deals with the different material types for reinforced hexagonal tubes, subject to quasi-static axial crushing. The third part of the experimental work consists of two primary steps; the first step is the fabrication of the specimens and the second step involves crushing tests.

### 7.2.1 Geometry and Mandrel Design

The hexagonal tube mandrels were made in a mechanical engineering laboratory at the Universiti Putra, Malaysia. The hexagonal section tubes had lengths of 100 mm while the thickness of the tubes depended on the material type. Table 7.1 gives the tube specifications for different types of materials. The apparatus used for tube fabrication is shown in Figure 7.1.

The three types of fiber reinforcement used to fabricate the composite hexagonal tubes were glass fiber yarn, kenaf fibers (mat and yarn), and carbon fibers (see Table 7.1). The images of these materials are shown in Figure 7.2. All fiber types were supplied by IBSB Innovate Pultrusion Sdn. Bhd Malaysia, D.E.R.™ 331™. Liquid epoxy glue and jointmine hardener were obtained from Dow Chemical Pacific Singapore via the Tazdiq Engineering Company.

**TABLE 7.1**
**The Material Specifications**

| Material | Density (g/cm$^3$) | Source |
|---|---|---|
| Carbon fiber | 1.6–1.7 | Supplier data sheet |
| Glass fiber | 2.5–2.59 | Yan et al. 2014 |
| Kenaf fiber | 0.18 | Abdewi et al. 2007 |

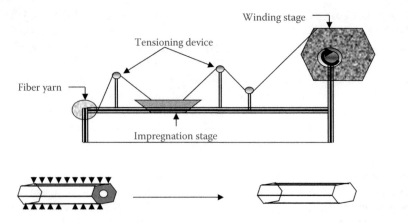

**FIGURE 7.1** Apparatus used for fabrication of the hexagonal composite tube's material.

**FIGURE 7.2** Types of fibers used in the current study. (a) Natural kenaf (mat). (b) Carbon fiber yarn. (c) Glass fiber yarn. (d) Natural kenaf yarn.

# Effects of Material Types on the Failure Modes 117

## 7.2.2 Fabrication Process

In this study, five cases of tube configuration were considered:

1. E-glass fiber: kenaf fiber mat/epoxy hexagonal tube termed GFKM
2. E-glass fiber: kenaf fiber yarn/epoxy hexagonal tube termed GFKY
3. Carbon fiber: kenaf fiber mat/epoxy hexagonal tube termed CFKM
4. Carbon fiber: kenaf fiber yarn/epoxy hexagonal tube termed CFKY
5. Kenaf fiber yarn: kenaf fiber mat/epoxy hexagonal tube termed KYKM

Table 7.2 summarises the description of these hexagonal composite tube specimens.

Many researchers have used a wet manual winding process in their studies to fabricate the composite structures (Yan et al. 2014; Abdewi et al. 2007). Accordingly, in this investigation, the hexagonal composite tubes were also achieved by a wet manual winding process, as shown in Figure 7.3.

The first step in the fabrication process was the preparation of the material and the mandrel. The surface of the mandrel was cleaned and dried, and a layer of nylon sheeting was rolled and applied around the solid wooden mandrel surface, with a layer of wax applied between the mandrel surfaces. This was to ensure that the composite structures could be easily extracted from the wooden mandrel surface.

In the second step, the tube was made by rolling the fibers onto a rotating mandrel of a suitable hexagonal section. During these steps, the epoxy and the hardener mixture should be applied to the fiber by hand, working it to avoid the formation of air bubbles between the layers of fibers. Reinforced fibers were continually wound onto a mandrel until the surface was covered and the required thickness was achieved. The third step of this process was the curing stage. The fabricated tube was cured at room temperature for 48 hours to obtain the optimum hardness and dryness. Finally, the tubes were cut into 100 mm lengths.

**TABLE 7.2**
**Description of the Hexagonal Composite Tube Specimens**

| Type of Tube | Hexagonal Angle ($\beta$) | Number of Layers | Height (mm) | Weight (Kg) |
|---|---|---|---|---|
| CFKM | 60 | 2 | 100 | 98 |
| GFKM | 60 | 2 | 100 | 100 |
| GFKY | 60 | 3 | 100 | 75 |
| KYKM | 60 | 2 | 100 | 115 |
| CFKY | 60 | 3 | 100 | 85 |

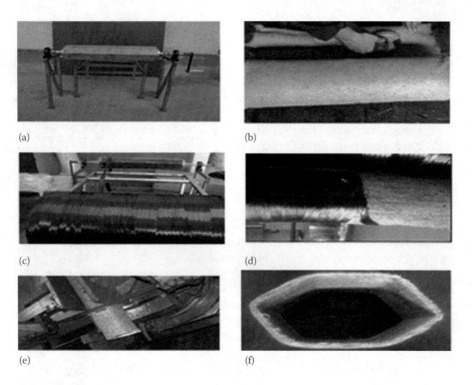

**FIGURE 7.3** Fabrication process. (a) Winding machine. (b) Preparing kenaf mat. (c) Winding processes. (d) Hybrid material. (e) Cutting stage. (f) Hexagonal tube ready for test.

## 7.3 CRUSH TESTING AND CRASHWORTHINESS PARAMETERS

In this study, the axial, quasi-static crushing tests were carried out using a universal testing machine (INSTRON MTS 3382) with a capacity load of 100 kN, at a constant speed of 15 mm/min. More details of the crushing test can be found in Alkbir et al. (2014). Crashworthiness is concerned with the energy absorption through controlled failure modes, which result in a gradual degeneracy in the load profile during energy absorption.

## 7.4 RESULTS AND DISCUSSION

### 7.4.1 Load-Displacement Curves and Failure Mechanisms

#### 7.4.1.1 Results of CFKM

Figure 7.4 shows the representative photographic images and the relationship between axial load and displacement. In the pre-crushing stage, the initial load amount that was accomplished was 60.099 kN at 5.2 mm displacement. The load

# Effects of Material Types on the Failure Modes

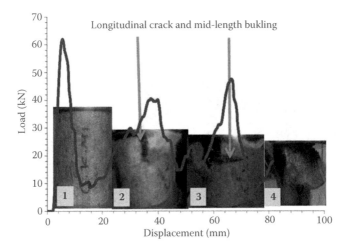

**FIGURE 7.4** Axial compression test of CFKM hexagonal tube.

value descended suddenly to a value of 10.68 kN at 11.8 mm. At this stage, the catastrophic failure mode was attributed to fiber cracking.

At the crushing stage, the tube initially resisted the compression load. This caused the load to increase progressively until it reached the second peak with a value of 39.165 kN at 36 mm. After this, the load dropped gradually and was associated with a longitudinal crack of the kenaf mat mode.

In the final stage, the carbon fiber was used to support a tube wall to stand against the upper metal plate movement, causing the load to increase again to the value of 45.9 kN as a densification stage occurred. In this case, the carbon shrank and did not break down (as shown in the picture). The same failure mechanism in the pre-crushing stage was demonstrated by Mahdi et al. (2014).

The failure modes observed for CFKM during axial compression tests were longitudinal crack and mid-length buckling in the outer layer in the tube.

### 7.4.1.2 Results of KYKM

The load-displacement curves for KYKM composite hexagonal tubes, 100 mm length when combined with kenaf (yarn and mat), were obtained from the axial crushing test. The tube behaved linearly in the pre-crushing stage. Figure 7.5 demonstrates how the load comes down due to the compression of the hexagonal tube. This substantial decrease in load is attributed to local buckling in mid-light tubes, with failure starting in the upper side of the composite hexagonal tube. Unstable behavior was noted at the post-crushing stage due to the catastrophic failure mode of the fibers. Oshkovr et al. (2012) used natural silk fibers to manufacture composite square tubes and reported that all the tubes crushed catastrophically at the pre-crushing stage.

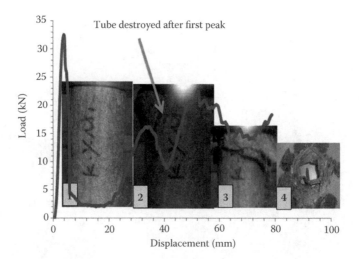

**FIGURE 7.5** Axial compression test of CFKM hexagonal tube.

### 7.4.1.3 Results of GFKY

On the load–displacement curve of E-glass fiber yarn and kenaf yarn (GFKY), the load increased in a linear manner until a first peak value of 12.992 kN at 5.07 mm displacement. Two crushing failure modes were observed immediately after this first peak—i.e., local bulking and transverse cracking (as shown in Figure 7.6). Stable behaviour was recorded due to the cohesion between the synthetic and natural fibers,

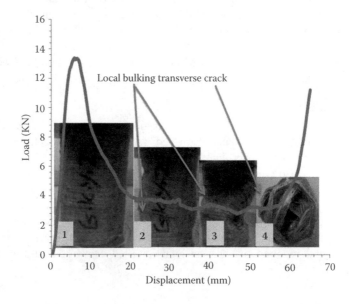

**FIGURE 7.6** Axial compression of the GFKY hexagonal tube.

# Effects of Material Types on the Failure Modes

with the crush energy of the hybrid material being absorbed by a chronological succession of various small cracking processes characterising every case of a crush. After a value of 40.88 kN, the load increased again, indicating that material densification had occurred. This phenomenon is presented (Mahdi et al. 2014) and can be explained by the effect of fiber orientation on the energy absorption capability of the axially crushed composite tubes.

### 7.4.1.4 Results of CFKY

The load–displacement curves for tubes with the carbon–kenaf yarn are displayed in Figure 7.7. It can be observed that this type of tube recorded the lowest value of initial load due to the brittleness of the kenaf yarn. The initial peak value was 6.16 kN at 9.99 mm displacement. After this, the load dropped slightly, before beginning to rise again (fluctuating around the mean crush load with a value of 5.5 kN at 58 mm displacement). This phenomenon was explained by the progressive folding and fragmentation of the tube wall in the post-crush stage, with the load increasing at the compaction zone until it reached 14.6 kN and 78 mm displacement. The types of failure mode observed and reported in this case were mode I, fragmentation of the tube, mode II unstable local buckling, and mid-length buckling.

### 7.4.1.5 Results of GFKM

The load–deformation curves for axially crushed hexagonal composite tubes reinforced by glass–kenaf mat GFKM are shown in Figure 7.8. As can be seen, the load firstly increased linearly with the displacement, up to the initial crush failure. After that, the load fell gradually and the tube became unable to resist the axial load. This coincided with the folding failure mode of the load, this being a progressive failure mode which fluctuated with increasing displacement. As the upper plate moved

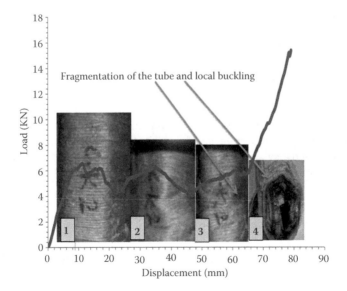

**FIGURE 7.7** Axial compression test CFKY hexagonal tube.

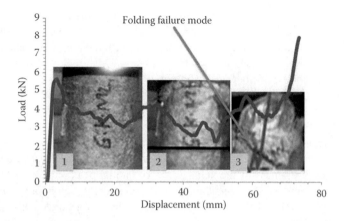

**FIGURE 7.8** Axial compression test of the (GFKM) hexagonal tube.

down, the tube began to compact, leading to a rapid increase in load in the final stage, where it was observed that the glass fiber competed without fiber crush (Yan et al. 2014), found the same failure mode in their investigation.

### 7.4.2 Results of Crashworthiness

The crashworthiness parameters of the axial crushing test of the hexagonal composite tubes were as follows:

- Initial peak load (Pi)
- Mean-crushing load (Pm)
- Crush force efficiency (CFE)
- Initial failure indictor (IFI)
- Specific energy absorption (Es)
- Specific energy absorption capability

Energy absorption capability during the structural crash is a requirement across the complete spectrum of passenger transportation. Instantaneous specific energy absorption capability of the segmented composite tube was computed. The total work done ($WT$) during the axial crushing of the tubes, which is equal to the area under the load–displacement curve is given by:

$$WT = \int_{Si}^{Sb} Pm \, ds \Rightarrow = Pm\,(Sb - Si) \tag{7.1}$$

where $WT$ is the total energy absorbed in the crushing of the composite tube specimens, and where $Sb$ and $Si$ are the crush distances.

Energy absorbed per unit mass (i.e., specific energy absorption) can calculate by substituting:

$$\text{SEA} = \frac{WT}{m} = \frac{Pm(sb-si)}{m} \text{ kJ/Kg} \tag{7.2}$$

where $m$ is the mass and $WT$ is the total energy absorbed in the crushing of the composite tube specimens. Various studies were defined specific energy absorption (SEA) as the energy absorbed per unit mass of material (Abosbaia et al. 2005; Mahdi et al. 2003; Eshkoor et al. 2013; Elgalai et al. 2004).

The relationship between specific energy and hybrid hexagonal composite tubes are shown in Figure 7.9. It is clearly seen that the tube made from carbon and kenaf mat (CFKM) shows the best overall performance in terms of energy absorption, whereas the other tube (including GFKM) shows the worst. The specific energy absorption values of the CFKM tubes (11.187 kJ/kg) are higher than those of KFYM (6.423 kJ/kg), CFKY (4.087 kJ/kg), GFKY (3.303 kJ/kg), and GFKM (2.350 kJ/kg). This phenomenon is due to the nature of the cohesion between the carbon fiber and the kenaf fiber mat. The specific energy absorption and the average crushing load of the hexagonal composite tubes. Here can be seen the difference between the specificenergy and average load of a tube reinforced with a carbon fiber–kenaf mat (CFKM) and tubes reinforced with kenaf mat only.

The CFKM specimen showed the highest peak load and specific energy load values when compared with the tube reinforced with the kenaf fibers. The differences between the crashworthiness parameters of the specimens are related to the usage of different types of material and relative improvements in specific energy absorption.

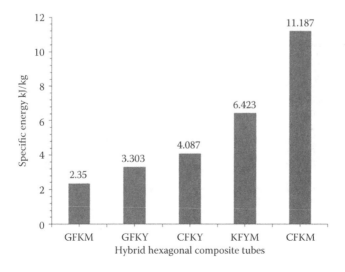

**FIGURE 7.9** Relationship between specific energy absorption and hybrid hexagonal tubes.

### 7.4.2.1 Effect of Hybrid Material in Initial Peak Load

The initial peak load value is the highest load that the tube specimen can reach before collapse occurs. Ataollahi et al. (2012) performed axial crushing testing for silk/epoxy composite square tubes. They reported that the length of tube walls of the tested specimens had a significant effect on the initial load value. Recently, Yan et al. (2014) investigated the lateral crushing of empty and polyurethane-foam-filled, natural flax fabric-reinforced epoxy composite tubes. They found that the increase in peak load capacity is almost directly proportional to the increase in the tube laminate layers. The first load is strongly dependent on the type of material reinforcement used. Figure 7.10, shows the initial capacity of the hexagonal composite tubes with different types of reinforcement (CFKY, GFKY, GFKM, KYKM, and Carbon fiber: kenaf fiber mat/epoxy hexagonal tube termed [CFMK]) under axial loading. The highest peak load value was achieved by CFKM with a value of 60.99 Kn, followed by KYKM with a value of 31.6 kN, GFKM with a value of 16.5 kN, GFKY with value 13 kN and CFKY with the value of 6.3 kN (Yan et al. 2014). Showed that the crushing behavior of the composite corrugated tube is sensitive to changes in the corrugation angle and the fiber type, which is strongly in agreement with the crush force efficiency (CFE).

One of the failure mode indicators is the crush force efficiency it is the ratio between the peak load and the average crushing load the crush force efficiency is the new parameter which introduced in this work, which helps to measure the tube cohesion-performance of an absorber (Meran 2015). The CFE is defined as:

$$(\text{CFE}). = \frac{Pi}{Pm}(\%) \qquad (7.3)$$

where $Pm$ is the average crushing load and $Pi$ the initial maximum peak load.

In hybrid fiber/epoxy composite hexagonal tubes, at an instant when the first tube crushed, the load immediately dropped significantly to the lowest value. Then, tubes

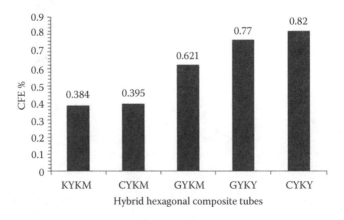

**FIGURE 7.10** Relationship between initial peak and hybrid hexagonal composite tubes.

Effects of Material Types on the Failure Modes 125

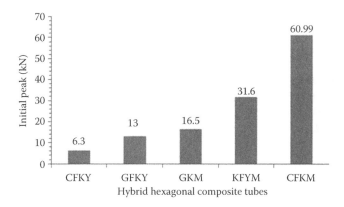

**FIGURE 7.11** Relationship between the crush force efficiency and hybrid hexagonal composite tubes.

stand and resisted the load applied. To present the structure cohesion so the value of CFE shows whether a unity or not, beside that failure mode can be considered such as progressive or catastrophic. Progressive deformation and stable collapse are desired features of vehicular structures; Figure 7.11 shows the CFE. For hybrid/epoxy composite hexagonal tubes, the results showed a decrease in the value of CFE from 0.37, 0.38, and 0.62 for GFKY, KYKM, and CFKM, respectively, which means the tube rang in unity. In general, the carbon/kenaf fiber (mat) hybrid fiber-reinforced hexagonal composites tubes CFKY specimens showed remarkably higher crush force efficiencies, meaning they had composite cohesion. Several attempts have been made to improve the mechanical properties of natural fibers. Amendment reinforcement materials and resin material and the manufacturing process are the main ways to improve the shortcomings. Hybridization with glass and carbon fibers provides a way to improve the mechanical properties of the natural fiber compounds for the application of structural automotive components.

### 7.4.2.2 Catastrophic Failure Mode Indicator

The catastrophic failure mode indicator (CFMI) is another significant measure of the failure mode that has been examined in previous study (Alkbir et al. 2014). This is defined as a criterion for determining whether the tubes crush in a catastrophic mode or undergo progressive failure.

When CFMI > 80%, the first crushing mode type is a catastrophic failure:

$$\text{CFMI} = \frac{d_m - d_i}{d_m - d_L} * 100 \tag{7.4}$$

where:

$d_m$: maximum displacement reachable
$d_i$: initial tube displacement
$d_L$: displacement load drop

It can be observed that a catastrophic failure mode indicator (CFMI) of 81% accrued in CFKM at the pre-crushing stage, whereas a 91% of CFMI appeared in KFYM at a catastrophic failure mode.

On the other hand, 79% of CFMI was found in CFKY material as a splaying failure mode. Moreover, GFKY material showed 66% CFMI associated with local buckling and transverse crack failure modes at the pre-crushing stage. Finally, GYKM material demonstrated a local buckling failure with catastrophic failure mode indicator of 57%.

## 7.5 CONCLUSIONS

The most appropriate compound materials for the fabrication of hexagonal composite tubes were successfully determined in the current work considering various hybrid materials. Due to environmentally friendly "green" composites based on nonwoven kenaf fiber mats and yarn, hybrid "synthetic" composites were created by incorporating fibrillated E-glass fiber yarn and carbon fiber yarn hexagonal tubes fabricated using modified epoxy resins. Their crashworthiness parameters including specific energy absorption, cash force efficiency, failure mode, and catastrophic failure mode were investigated. Based on the results obtained, the following conclusions are drawn from this study:

1. Hexagonal composite tubes from carbon fiber yarn and non-woven kenaf natural fibers exhibited good energy absorption capability as well as a stable load-carrying capacity. The specific energy absorption of CFKM significantly improved from 11.18 kJ/kg compared with CFKY with the value of 4.08 kJ/kg, GFKY with the value of 3.30 kJ/kg, and GFKM with the value of 2.35 kJ/kg.
2. High initial failure crush load lead to a catastrophic failure mode as well as an unstable post-crush stage.
3. The results showed a decrease in the value of (CFE for hybrid/epoxy composite hexagonal tubes.) from 0.37, 0.38, and 0.62 for GFKY, KYKM, and CFKM, respectively, which meant the tube rang in unity.
4. Although the axially loaded hexagonal tube-reinforced carbon yarn and non-woven kenaf/epoxy specimens were stronger and stiffer than the others, their crush behavior was unstable at cushing stage.
5. Catastrophic failure mode, splaying failure mode, local buckling, and transverse crack failure mode were observed and recorded.

## 7.6 FUNDING

The financial support from Universiti Putra Malaysia through the Research, University Grant Scheme vote no. 9301200 is highly appreciated.

## REFERENCES

Abdewi, E.F. (2007). Energy Absorption Characteristics of Radially Corrugated Composite Shells Under Different Quasi-static Loading Conditions (Doctoral dissertation, Universiti Putra Malaysia).

Abdewi, E.F., S. Sulaiman, A.M.S. Hamouda, and E. Mahdi. 2006. Effect of geometry on the crushing behaviour of laminated corrugated composite tubes. *J. Mater. Process. Technol.* 172, 394–399.

Abdewi, E.F., S. Sulaiman, A.M.S. Hamouda, and E. Mahdi. 2008. Quasi-static axial and lateral crushing of radial corrugated composite tubes. *Thin-Walled Struct.* 46, 320–332.

Abosbaia, A.A.S., E. Mahdi, A.M.S. Hamouda, and B.B. Sahari. 2003. Quasi-static axial crushing of segmented and non-segmented composite tubes. *Compos. Struct.* 60, 327–343.

Abosbaia, A.S., E. Mahdi, A.M.S. Hamouda, B.B. Sahari, and A.S. Mokhtar. 2005. Energy absorption capability of laterally loaded segmented composite tubes. *Compos. Struct.* 70, 356–373.

Alkbir, A.A. Nurain, S. Materials, D.M. Models et al. 2016. International Journal of Effect of crashworthiness parameters in natural fibre-reinforced polymer composite tubes: A literature review. *Composite Structure*.

Alkbir, M.F.M., S.M. Sapuan, A.A. Nuraini, and M.R. Ishak. 2014. Effect of geometry on crashworthiness parameters of natural kenaf fibre reinforced composite hexagonal tubes, *Mater. Des.* 60, 85–93.

Ataollahi, S., S.T. Taher, R.A. Eshkoor, A.K. Ariffin, and C.H. Azhari. 2012. Energy absorption and failure response of silk/epoxy composite square tubes: Experimental. *Compos. Part B Eng.* 43, 542–548.

Dittenber, D.B. and H.V.S. Gangarao, 2012. Critical review of recent publications on use of natural composites in infrastructure. *Compos. Part A Appl. Sci. Manuf.* 43, 1419–1429.

Elgalai, A.M., E. Mahdi, A.M.S. Hamouda, and B.S. Sahari. (2004). Crushing response of composite corrugated tubes to quasi-static axial loading. *Compos. Struct.* 66, 665–671.

El-Shekeil, Y.A., S.M. Sapuan, M. Jawaid, O.M. Al-Shuja'. 2014. Influence of fiber content on mechanical, morphological and thermal properties of kenaf fibers reinforced poly(vinyl chloride)/thermoplastic polyurethane poly-blend composites. *Mater. Des.* 58, 130–135.

Eshkoor, R.A., S.A. Oshkovr, A.B. Sulong, R. Zulkifli, A.K. Ariffin, and C.H. Azhari. 2013. Effect of trigger configuration on the crashworthiness characteristics of natural silk epoxy composite tubes. *Compos. Part B Eng.* 55, 5–10.

Eshkoor, R.A., S.A. Oshkovr, A.B. Sulong, R. Zulkifli, A.K. Ariffin, and C.H. Azhari. 2013. Comparative research on the crashworthiness characteristics of woven natural silk/epoxy composite tubes. *Mater. Des.* 47, 248–257.

Hull, D. 1991. A Unified Approach to Progressive Crushing of Fiber-Reinforced Composite Tubes. *Compos. Sci. Technol.* 40, 377–421.

Maepa, C.E., J. Jayaramudu, J.O. Okonkwo, S.S. Ray, E.R. Sadiku, and J. Ramontja. 2015. Extraction and Characterization of Natural Cellulose Fibers from Maize Tassel. *Int. J. Polym. Anal. Charact.* 20, 99–109.

Mahdi, E., A.M.S. Hamouda, B.B. Sahari, and Y.A. Khalid. 2003. Effect of hybridisation on crushing behaviour of carbon/glass fibre/epoxy circular–cylindrical shells. *Journal of Materials Processing Technology.* 132, 49–57.

Mahdi, E. and T.A. Sebaey. 2014. An experimental investigation into crushing behavior of radially stiffened GFRP composite tubes. *Thin-Walled Struct.* 76, 8–13.

Mamalis, A.G., D.E. Manolakos, M.B. Ioannidis, and D.P. Papapostolou. 2004. Crashworthy characteristics of axially statically compressed thin-walled square CFRP composite tubes: Experimental. *Compos. Struct.* 63, 347–360.

Melo, J.D.D., A.L.S. Silva, and J.E.N. Villena. 2008. The effect of processing conditions on the energy absorption capability of composite tubes. *Compos. Struct.* 82, 622–628.

Meran, A.P. 2015. Solidity effect on crashworthiness characteristics of thin-walled tubes having various cross-sectional shapes. *Int. J. Crashworthiness*. 8265, 1–13.

Meredith, J., R. Ebsworth, S.R. Coles, B.M. Wood, and K. Kirwan. 2012. Natural fibre composite energy absorption structures. *Compos. Sci. Technol.* 72, 211–217.

Mohanty, a. K., M. Misra, and L.T. Drzal. 2001. Surface modifications of natural fibers and performance of the resulting biocomposites: An overview. *Compos. Interfaces*. 8, 313–343.

Oshkovr, S.A., R.A. Eshkoor, S.T. Taher, A.K. Ariffin, and C.H. Azhari. 2012. Crashworthiness characteristics investigation of silk/epoxy composite square tubes. *Compos. Struct*. 94, 2337–2342.

Palanivelu, S., W. van Paepegem, J. Degrieck, J. van Ackeren, D. Kakogiannis, and D. Van. 2010. Experimental study on the axial crushing behaviour of pultruded composite tubes. *Polym. Test.* 29, 224–234.

Ramesh, M., K. Palanikumar, and K.H. Reddy. 2013. Mechanical property evaluation of sisal-jute-glass fiber reinforced polyester composites. *Compos. Part B Eng.* 48, 1–9.

Saheb, D.N., and J. P. Jog. 1999. Natural Fiber Polymer Composites: A Review. *Adv. Polym. Technol*. 18, 351–363.

Sapuan, S.M., F. Pua, Y. a. El-Shekeil, and F.M. AL-Oqla. 2013. Mechanical properties of soil buried kenaf fibre reinforced thermoplastic polyurethane composites. *Mater. Des.* 50, 467–470.

Shibata, S., Y. Cao, and I. Fukumoto. 2006. Lightweight laminate composites made from kenaf and polypropylene fibres. *Polym. Test.* 25, 142–148.

Terborgh, J. 1974. Preservation of natural diversity, *BioScience* 24, 715–722.

Warrior, N.A., T.A. Turner, E. Cooper, and M. Ribeaux. 2008. Effects of boundary conditions on the energy absorption of thin-walled polymer composite tubes under axial crushing. *Thin-Walled Struct.* 46, 905–913.

Yan, L., N. Chouw, and K. Jayaraman. 2014. Effect of triggering and polyurethane foam-filler on axial crushing of natural flax/epoxy composite tubes. *Mater. Des.* 56, 528–541.

Yan, L., N. Chouw, and K. Jayaraman. 2014. Flax fibre and its composites—A review. *Compos. Part B Eng.* 56, 296–317.

Yan, L., N. Chouw, and K. Jayaraman. 2014. Lateral crushing of empty and polyurethane-foam filled natural flax fabric reinforced epoxy composite tubes. *Compos Part B* 63 (2014), 15–26.

Yousif, B.F., A. Shalwan, C.W. Chin, and K.C. Ming. 2012. Flexural properties of treated and untreated kenaf/epoxy composites. *Mater. Des.* 40, 378–385.

# 8 Eco-Friendly Kenaf Hybrid Materials

*S. Norshahida and H. Ismail*

## CONTENTS

8.1 Introduction .................................................................................................. 129
8.2 Experiment .................................................................................................... 131
    8.2.1 Materials ........................................................................................... 131
    8.2.2 Sample Fabrication ........................................................................... 131
    8.2.3 Outdoor Weathering Test .................................................................. 132
    8.2.4 Characterizations .............................................................................. 132
8.3 Results and Discussion ................................................................................. 133
    8.3.1 Physical Appearance of LDPE/TPS/KCF/HC Composites after Exposed to Natural Weathering .................................................. 133
    8.3.2 Tensile Properties of LDPE/TPS/KCF/HC Composites after Exposed to Outdoor Natural Weathering ................................... 133
    8.3.3 FTIR Analysis of LDPE/TPS/KCF/HC Composites after Exposed to Outdoor Natural Weathering ................................... 138
    8.3.4 Morphological of LDPE/TPS/KCF/HC Composites after Exposed to Outdoor Natural Weathering ................................... 139
    8.3.5 FTIR Analysis of LDPE/TPS/KCF/HC Composites after Exposed to Outdoor Natural Weathering ................................... 140
8.4 Concluding Remarks ..................................................................................... 141
References ............................................................................................................ 142

## 8.1 INTRODUCTION

Petroleum-based plastics have been focus of attention as they exhibit promising advantages and have replaced conventional materials (glass, ceramics, and metals). They have been used with other materials in many applications in regard to their low cost, especially in processing, as well as their outstanding performance. Their production and usage have been tremendously increased owing to their lightweight properties, durability, and adaptability to desired forms (Rahmat et al. 2009).

Despite their attractive performance, they are hardly degradable and these may lead to many environmental problems associated with their disposal. Taking into account, these serious issues have led to the exploration of new types of materials which pose alternatives to petroleum-based plastics. Recent research has focused on the utilization of natural resources in place of the synthetic ones so that they can be degraded in selective environments. Therefore, this may be considered as one of many solutions to waste disposal problems.

Numerous studies have been focused on improving the mechanical properties while, at the same time, retaining the biodegradability of composites by blending synthetic polymers with some natural materials, including biopolymers, fibers, and inorganic mineral fillers. Biodegradable polymers or so-called biopolymers are predominantly those from renewable resources, particularly polymers obtained from agro resources, such as starch, that have long been recognized. The strategy is that biodegradable polymers (i.e., starch) are introduced into conventional plastics, which then promote the accessibility of the plastics to oxygen and microorganisms. This has guaranteed at least partial degradation (He et al. 2012). The major sources of starches that commonly used for production of biodegradable polymers are maize, tapioca, and rice. However, the properties of sago starch blends are rarely reported. Indeed, Malaysia is well known as a country with an enormous cultivation of the sago palm. Thus, with concerns to the current needs of developing degradable materials makes sago starch as an attractive and promising material for blending with synthetic polymers (i.e., polyethylene). However, it has been found that the addition of minor content of biodegradable polymers in the synthetic polymer causes a reduction of the mechanical properties due to the immiscibility caused by the different polarities of the two constituents (Prachayawarakorn et al. 2010).

Therefore, approaches have been undertaken to explore the effectiveness of using a natural fiber as a reinforcing agent in polymer matrices (Mukherjee and Kao 2011). Under certain conditions and periods, natural fiber-based products have proved their biodegradability which proposes them as a reliable approach for technically degradable composites. Among the various natural fibers, kenaf fibers are of particular interest. Kenaf attracts a special attention due to its good mechanical properties and its increasing cultivation since it grows faster and can be harvested at a very low cost.

Apart from the noteworthy points, one of the main challenges related to the use of natural fibers in composites is their high moisture absorption, which subsequently leads to poor interfacial bonding with matrices (Khan et al. 2010). At present, the resolution does not only rely on the chemical modification of the fibers. With the recent advancement in technologies, much attention has been paid to the development of composites in which two or more fillers are utilized in a matrix to obtain diverse properties in the composites (Muhammad Safwan et al. 2013). Hybrid composites show encouraging results as they offer a balance between performance properties. Thus, recently, organic/inorganic materials filled composites were widely researched. Inorganic fillers have drawn a huge interest because of their cost efficiency and abundant availability. Various thermoplastic composites prepared by the inclusion of inorganic fillers such as montmorillonite, mica, calcium carbonate, and carbon nanotubes have demonstrated outstanding improvements in mechanical and thermal properties, dimensional stability, gas permeability, physicochemical behaviors, as well as biodegradability in comparison with neat thermoplastics (Nakamura et al. 2013). In recent years, halloysite clays (HCs) have been incorporated as a new type of reinforcing filler in polymers (Ismail and Shaari 2010). Halloysite seems to be promising as it can be easily dispersed in a polymer matrix by shearing owing their rod like geometry and limited intertubular contact area.

Since this type of polymer composites has gained interest and staple of industry, their durability under various environmental conditions and degradability after service life is an unavoidable fact. Earlier, it has been reported that nanoparticle-filled composites degrade faster than the pristine polymers because of the decomposition of ammonium ion which create acidic sites on layered silicates. This can induce the formation of free radicals upon UV irradiation which will then lead the materials to degradation (Kumar et al. 2009). Yet, very limited works are available in open literature on the durability or interaction of hybrid materials with environmental factors. Thus, it is a good opportunity to explore the potential use of natural materials and hybrid structures with an overview on the processability, compatibility, and degradation properties, focusing on the aspects that most relevant to the preferred composites.

## 8.2 EXPERIMENT

### 8.2.1 MATERIALS

Sago starch (13% moisture) was obtained from the Land Custody Development Authority (LCDA), Sarawak, Malaysia. It had an average particle size of 20 μm and decomposition temperature of 230°C. Glycerol (plasticizer), an analytical grade reagent, was purchased from Merck Chemicals (Malaysia) and used as received. Low density polyethylene (LDPE, LDF 260GG) with melt flow index of 5 g/10 min was obtained from Titan (M) Sdn. Bhd. (Malaysia). Kenaf fibers (core) with average length of 5 mm was supplied by National Kenaf and Tobacco Board (LKTN), Malaysia. The fibers were subjected to grinding process which yields particles approximately 70–250 μm in diameter. KCF were then dried for 3 hrs at 70°C using vacuum before being used in the subsequent composite fabrication. The ultrafine grade halloysite clay was supplied by Imerys Tableware Asia Limited, New Zealand. The density of halloysite clay is 2.14 g/cm$^3$ (Pasbakhsh et al. 2009). Typical dimensions of HC; 150 nm up to 2 μm long with 20–100 nm outer diameter and 5–30 nm inner diameters (Ismail et al. 2008).

### 8.2.2 SAMPLE FABRICATION

Sago starch powder was vacuum dried by heating at 80°C for 24 hrs before blending and processing. The dried sago starch was then premixed with glycerol by using a high speed mixer. The weight ratio of sago starch and glycerol was maintained at 65:35% (wt). The blend was stored overnight to allow the diffusion of glycerol into starch granules which would help the melt mixing process. Inorganic mineral filler (halloysite clay) in powder form was dried at 105°C in a convection oven for 24 hrs to expel moisture prior to use. The ratio between LDPE/TPSS was fixed at 90–10% (wt) and the KCFs were constantly loaded at 10 phr. For this work, inorganic filler loading varied from 3 to 15 phr. Mixing was carried out using an internal mixer Polydrive Thermo Haake R600. The mixing procedures involved melt-blended of the samples at 150°C at a speed of 50 rpm for a period of 20 min. The processed samples were then compression-molded in an electrically heated hydraulic press (Kao Tieh Go Tech Compression Machine) at 50°C into a 1-mm thick sheet.

### 8.2.3 Outdoor Weathering Test

The tests were performed by exposing dumbbell samples of LDPE/TPSS blends filled with KCF and halloysite clay composites to the outdoor weathering. The tests were conducted in an open area at Universiti Sains Malaysia (USM), Penang, Malaysia (latitude 5°28′N, longitude 100°29′E) for a period of 3 months until 6 months. The dumbbell samples were arranged on an exposure rack facing to the south and at an inclination angle of 45°. The experiment set up is shown in Figure 8.1. After the exposed period, the samples were subjected to further mechanical and analytical tests. The collected samples were washed with distilled water, dried, and weighed after drying to a constant weight in an air-drying oven maintained at 70°C.

### 8.2.4 Characterizations

Tensile tests were carried out with a universal testing machine (Instron 3366) according to ASTM D638. A crosshead speed of 5 mm/min was used and the test was performed at temperature of $25 \pm 3°C$ and relative humidity of $60 \pm 5\%$. Five specimens were used to obtain average values for tensile strength, elongation at break, and Young's modulus. The retention of these properties was calculated using the following equation (Equation 8.1):

$$\text{Retention}(\%) = \frac{\text{Value after degradation}}{\text{Value before degradation}} \times 100 \qquad (8.1)$$

**FIGURE 8.1** Outdoor natural weathering experimental set up.

The functional groups and chemical characteristics of composites before and after natural weathering test as a function of halloysite clay content were obtained by Fourier Transform Infrared Spectroscopy (FTIR, Perkin Elmer System 2000) with a resolution of 4 cm$^{-1}$ in a spectral range of 4000–600 cm$^{-1}$ using 32 scans per sample. Carbonyl index (CI) was used as a parameter to observe the degree of degradation. CI was calculated as the ratio of the intensity of the peak at 1720 cm$^{-1}$ to the reference peak intensity of 1460 cm$^{-1}$. The peak at 1720 cm$^{-1}$ corresponded to the absorption from the presence of carbonyl groups, which was the by-product of polymer degradation.

Scanning electron micrographs of surfaces of the composites before and after outdoor weathering were obtained by using a scanning electron microscope (SEM, model ZEISS Supra 35 VP). The samples were sputter coated with a thin layer of carbon to avoid electrostatic charging during the examination.

The biodegradability was assessed and evaluated by measuring weight loss before and after testing. The weight loss of weathered samples was calculated and evaluated using the following Equation 8.2:

$$\text{Weight Loss}(\%) = \frac{W_1 - W_0}{W_0} \times 100 \tag{8.2}$$

Where $W_0$ and $W_1$ are sample weights before and after the composting test, respectively.

## 8.3 RESULTS AND DISCUSSION

### 8.3.1 Physical Appearance of LDPE/TPS/KCF/HC Composites after Exposed to Natural Weathering

Figure 8.2 displays a series of samples retrieved after exposure to natural weathering for 6 months. The samples consisted of LDPE/TPSS blends with the addition of KCF/HCs hybrid fillers as a function of HC loading. From the figure, it can be seen, the samples showed discoloration after exposure to weathering. Samples that were yellowish in color originally, turned out to be grayish upon higher concentration of HCs loaded in the composites as well as those under a longer exposure time. This mainly resulted from the deterioration upon the combined effect of water and UV light. During weathering, the absorption of water by polar components (namely TPSS, KCF, as well as HCs) might have caused the composites to be more accessible to UV light penetration and oxidation. This somehow resulted in discoloration of the composite strips after exposure (Eshraghi et al. 2013).

### 8.3.2 Tensile Properties of LDPE/TPS/KCF/HC Composites after Exposed to Outdoor Natural Weathering

Figures 8.3–8.5 llustrate the tensile properties of LDPE/TPSS blends reinforced with hybrid fillers (KCF/HC) as a function of HC loading before and after exposure to outdoor natural weathering. The extent of degradation experienced by the composites

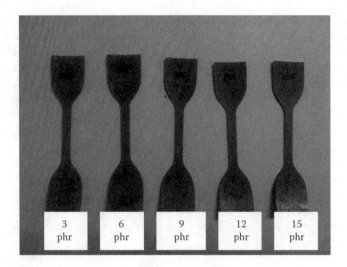

**FIGURE 8.2** Visual observation of composites with the addition of halloysite clays at various loadings after outdoor natural weathering test for 6 months.

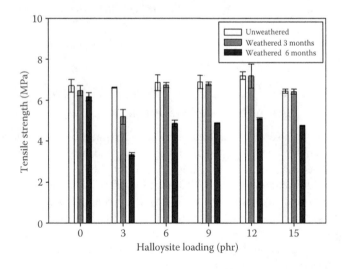

**FIGURE 8.3** Tensile strength of composites with the addition of halloysite clay before and after exposure to outdoor natural weathering for 3 and 6 months.

upon 3 and 6 months of exposure was carefully measured. Referring to Figure 8.3, it can be observed that the unweathered samples showed a gradual increase of tensile strength with the addition of HCs into composites and the optimum loading was obtained at 12 phr. Further loading of HCs caused an abrupt decrement of strength.

For weathered samples, several important observations can be addressed. It seems that the weathered samples exhibited similar tendency with unweathered ones,

# Eco-Friendly Kenaf Hybrid Materials

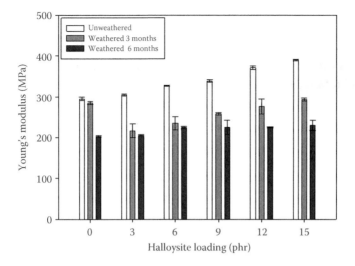

**FIGURE 8.4** Young's modulus of composites with the addition of halloysite clay before and after exposure to outdoor natural weathering for 3 and 6 months.

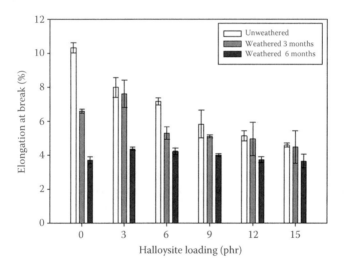

**FIGURE 8.5** Elongation at break of composites with the addition of halloysite clay before and after exposure to outdoor natural weathering for 3 and 6 months.

whereby, an increase HC loadings tended to augment the tensile strength until an optimum point. Yet, a decay of strength values still could have been obtained upon degradation of the weathered samples. In fact, prolonged exposure time resulted in the progressive decline of strength. The retention values in Table 8.1 imply the degradation rate. It is interesting to note that the samples without the addition of HCs (control samples) experienced a slow degradation process, whereby the

## TABLE 8.1
Retention Ratios of LDPE/TPSS Blend Reinforced KCF/HC Hybrid Fillers at Different HC Loading after Exposure to Natural Weathering for 3 and 6 Months

| | Retention (%) | | | | | |
|---|---|---|---|---|---|---|
| | Tensile Strength | | Elongation at Break | | Young's Modulus | |
| HC Loading (phr) | 3 Months | 6 Months | 3 Months | 6 Months | 3 Months | 6 Months |
| 0 | 99.96 | 96.44 | 63.73 | 35.82 | 89.98 | 68.84 |
| 3 | 78.61 | 50.61 | 95.26 | 54.66 | 71.18 | 67.40 |
| 6 | 98.34 | 70.93 | 73.80 | 59.08 | 71.84 | 68.63 |
| 9 | 98.72 | 70.75 | 87.42 | 68.56 | 76.19 | 66.35 |
| 12 | 99.76 | 70.86 | 96.70 | 72.84 | 74.30 | 60.61 |
| 15 | 99.61 | 73.76 | 106.65 | 79.63 | 75.31 | 58.99 |

tensile strengths are declining to only 3.64% and 7.90%, after being weathered for 3 and 6 months, respectively. On the other hand, samples loaded with HCs results a significant decay in tensile strength upon 3 months of exposure, and the strengths seems to drop drastically after 6 months of weathering.

In all cases, the strength of the composites showed a decline with a prolonged weathering time up to 6 months, most probably due to the combination of degrading effect on the composites during exposure; photodegradation, temperature/moisture effects, hydrolysis as well as microbial attacks (Ahmad Thirmizir et al. 2011). It is believed that, upon weathering, there is occurrence of the photooxidation process, which basically degraded the polymer matrix (LDPE) by breaking down of polymer chains. The degradation proceeded according to Norrish type I and II reactions which lead to chain scission and the formation of carbonyl groups. As a result, the molecular weight and mechanical properties of the polymer had weakened (Darie et al. 2011). It is worth noting that, with the addition of starch into the synthetic polymer, the blend is more prone to degradation. Exposure of starch to UV radiation leads to degradation by photooxidation of both components of starch, amylose, and amylopectin (Pang et al. 2013).

Concurrently, when the fiber-reinforced polymer composites are exposed to outdoor weathering, fiber components such as lignin, hemicelluloses and cellulose also underwent rapid degradation. During natural weathering, the absorption of UV lights by lignocellulosic components lead to photodegradation along with moisture, temperature and oxidative agents such as oxygen, which caused the breakdown of lignin and cellulose components. Accordingly, some physical, chemical, and biological properties were diminished (Ahmad Thirmizir et al. 2011).

Apart from the aforementioned degradation mechanisms for TPSS, KCF, and LDPE, the possible explanation for this occurrence is that the nanoparticles like HCs might accelerate the degradation reaction by acting as catalysts for the photooxidation of LDPE (Eshraghi et al. 2013). The transition metal ions are likely inducing oxidation of LDPE and eventually act as catalysts for decomposition of hydroperoxide (by reducing hydroperoxide into hydroxyl radicals) (Kumanayaka et al. 2010). A similar assumption was postulated by Botta et al. (2009) and Bussiere et al. (2013). Another possible theory as to the cause of detrimental effects to the photooxidative stability of polymers is the generation of acidic sites on the layered silicates due to the decomposition of ammonium ions (Kumanayaka et al. 2010). These active sites will then accept an electron from the donor molecules of the polymer matrix and formed free radicals upon UV radiation. These will lead to the oxidation and break of molecular chains. Thus, the material suffers degradation and eventually decreases the strength (Kumar et al. 2009). It is well-known that the inclusion of clay responsible for greater degradation rate and critical effect observed on LDPE during UV irradiation. As such, FTIR analysis shown in later sections ascribed the formation of degradation product associated with the collapse of LDPE by photooxidation. It has been reported by Eshraghi et al. (2013) whereby the addition of $SiO_2$ and organically modified MMT to composites have resulted an appearance of new peaks correspond to carbonyl, vinyl and hydroxyl groups during UV degradation. Apart from that, it is believed that the hybridization of KCF and HCs particularly at higher HC loading might result in a less dense structure due to agglomeration effect. By some means, the formation of voids/pores facilitated the degradation process considering the weathering factors such as water absorption, UV light penetration as well as microorganism's attack allow the deterioration of TPSS, KCF, HCs, and subsequently LDPE.

Likewise, Young's modulus of unweathered samples rises gradually as more HCs loaded in the composites (3–15 phr). It is well-known the incorporation of nanofillers (rigid filler particles) will stiffen the composites and lead to augment of the modulus. A similar tendency was observed for weathered samples (as shown in Figure 8.4). However, the retention values shown in Table 8.1 decreased after exposure to weathering for 3 months and declined further upon prolonged exposure time (6 months). This could be attributed to the degradation mechanism mentioned earlier: thermal degradation, hydrolysis, photodegradation, and microorganism's attack which somehow lead to the formation of pores and cracks on the composite. This correlated to the SEM morphology of the composite surface shown in the later section. Prolonged exposure time up to 6 months allowed for a longer degradation process to take place and the composites deteriorated more. This rendered the ductility of polymer brittle and consequently dropped the Young's modulus.

Referring to Figure 8.5, it has been found that the elongation at break declined with the inclusion of HCs in composites. The decrement is predicted, considering the polymer chain mobility and deformability is hindered by the presence of HCs. Correspondingly, after weathering exposure, the elongation at break also declined as the HC loading increased. This was confirmed by the retention values for 3 months and 6 months weathered composite samples. The decline in elongation at break with an increase in the weathering time was due to the embrittlement of the overall composite system associated with degradation. Upon weathering, the composite samples underwent extensive chain scissions which broke down tie chain molecules and entanglements. Thus, the ductility of polymer was impaired and the elongation at break dropped subsequently.

### 8.3.3 FTIR ANALYSIS OF LDPE/TPS/KCF/HC COMPOSITES AFTER EXPOSED TO OUTDOOR NATURAL WEATHERING

Figure 8.6 demonstrates the representative IR spectrum of LDPE/TPSS/KCF composites with the addition of 12 phr HCs obtained before and after exposure to natural weathering. From the figure, it can be seen that before weathering, the samples exhibited typical characteristic peaks at 2913 cm$^{-1}$ and 2847 cm$^{-1}$, 1474 cm$^{-1}$, and 722 cm$^{-1}$ corresponding to stretching vibrations of C-H stretching, -CH$_3$ bending, and –CH$_2$ vibration, respectively. In addition, the absorption bands around 919 cm$^{-1}$ and 1043 cm$^{-1}$ are associated with the Al-OH vibrations, and Si-O stretching bands can be observed (Pasbakhsh et al. 2010; Alhuthali and Low 2013).

After 3 months of weathering (t = 3), it seemed that the samples exhibited typical IR spectra, except for some changes of intensity and appearance of some peaks. As weathering progressed, the polymer underwent photooxidation and was facilitated by the presence of HCs. Accordingly, the broad peak observed around the wavenumber of 3000–3600 cm$^{-1}$ corresponded to the formation of hydroxyls, which was an indication of UV degradation products (Kumanayaka et al. 2010). This was evidenced by the close examination of IR spectra in region (i).

Meanwhile, carbonyl groups (C=O) exhibited the most significant changes correspondeding to the oxidation and degradation of the composite system during the period of outdoor exposure. After exposure to weathering, there is an appearance of

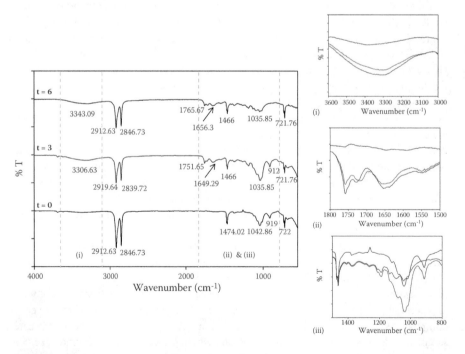

**FIGURE 8.6** Representative FTIR spectra of composites with the addition of 12 phr halloysite clay before and after exposure to outdoor natural weathering for 3 and 6 months.

carbonyl groups in the range of 1800–1600 cm$^{-1}$. A close examination of IR spectra in region (ii) showed that the peaks are relatively broad which signify functional groups within carbonyl family, namely ketones, carboxylic acids, and vinyl groups (Eshraghi et al. 2013). Accordingly, it was assumed that the presence of HC affected the photooxidation reaction. This peak intensified continuously with prolonged exposure time. In this case, the carbonyl index (CI) value was negligible because of the relatively wide range of carbonyl region.

In addition, there was a decreased intensity of peak at the band of 1043 and 919 cm$^{-1}$ observed in the IR spectrum of weathered samples (as shown in region [iii]). It was reported that the transition metal ions are likely induce oxidation of LDPE and eventually act as catalyst for decomposition of hydroperoxide (by reducing hydroperoxide into hydroxyl radicals. Likewise, the generation of acidic sites on the layered silicates due to the decomposition of ammonium ions also resulted in detrimental effects to the photooxidative stability of polymers (Kumanayaka et al. 2010). Thus, the decrement of intensities was probably ascribed to extensive chain scissions which break down ties chain molecules and entanglements due to photodegradation, and was also an indication of embrittlement of the polymer upon weathering (Dash et al. 2000). Prolonged weathering time allowed the intensities to decrease further or disappear. Clearly, the extent of photodecomposition reactions was indicated by the intensities of these peaks.

### 8.3.4 Morphological of LDPE/TPS/KCF/HC Composites after Exposed to Outdoor Natural Weathering

Figure 8.7 illustrates the morphology of LDPE/TPSS blends reinforced KCF with the addition of HC (at various loading) after exposure to natural weathering for 3 and 6 months. After 3 months of exposure, composite samples with 3 phr HC loading showed a rougher surface with some fungal colonies (Figure 8.7a). As more HCs loaded in the composites (12 phr), the pores and crack of different sizes and shapes are obviously seen on the surface (Figure 8.7b). The formation of pores and cracks indicated that the composites underwent significant thermal degradation, hydrolysis, photodegradation, and microorganism attacks upon weathering. It was believed that the degradation of individual components in the composites contributed to the deterioration in mechanical properties as well as morphology. In this case particularly, it seems that the higher degradation rate is reflected by the role of HCs which accelerated the reaction by acting as catalyst in the photooxidation of LDPE. Apart from that, the photooxidation reaction was also induced by the decomposition of ammonium ions which then created acidic sites on the layered silicates (Kumanayaka et al. 2010). These demonstrated the extent of degradation suffered by composites upon weathering accordingly. For that reason, composite samples with a higher loading of HCs (15 phr) lead to severe degradation described by increasing the number and size of pores and cracks shown in Figure 8.7c.

The degradation effects seen on the surface of the composites were more prominent after 6 months exposure to natural weathering. For composites with low content of HCs (3 phr), surface crazing is evidently seen throughout the surface (as shown in Figure 8.7d). This meant that the polymer underwent cyclic expansion and

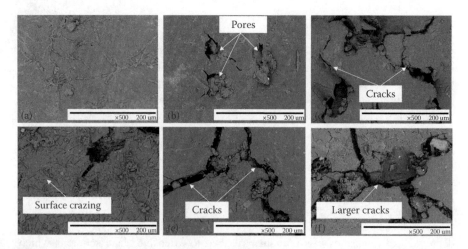

**FIGURE 8.7** SEM morphology of LDPE/TPSS blends reinforced KCF with the addition of halloysite clay after exposure to outdoor natural weathering for 3 and 6 months: (a) 3 phr HC–3 months; (b) 12 phr HC–3 months; (c) 15 phr HC–3 months; (d) 3 phr HC–6 months; (e) 12 phr HC–6 months; (f) 15 phr HC–6 months (500 × magnification).

contraction as well as the photodegradation process. The structural damages are reflected from the embrittlement of composites upon degradation. As observed in Figure 8.7e, the severity of cracks increased as more HCs were loaded into composites (12 phr) and the samples were exposed to prolonged weathering. As mentioned in earlier discussions, the presence of HCs clearly sped up the rate of photooxidation and overall degradation, which consequently left larger cracks. As the HCs content increased further up to 15 phr, the failure seems more evident (Figure 8.7f). This was because the nanofillers tended to agglomerate and act as a stress concentration point. Eventually, the cracks initiated brittle fracture and become worse upon photooxidation reactions. The cracks allowed deeper microbial invasion and impaired the overall composite structure. This was further justified by the measured weight loss.

### 8.3.5 FTIR ANALYSIS OF LDPE/TPS/KCF/HC COMPOSITES AFTER EXPOSED TO OUTDOOR NATURAL WEATHERING

The changes in weight for the weathered samples are shown in Figure 8.8. After 3 months of exposure to natural weathering, it can be seen that the samples suffered a slight weight loss as HC loading increase. The percentage of weight loss was found to be 4.13–4.51% when 3 to 15 phr HCs loaded in the composites, respectively. The prolonged exposure time was up to 6 months, whereas results showed a linear increase of weight loss with the addition of HCs from 3 phr up to 15 phr. Likewise, the percentages of weight loss were significantly higher for 6 months of exposure compared to the one with a shorter weathering period. These were by some means reflected in the tensile properties and morphology obtained in previous sections.

On the basis of the results obtained, several assumptions can be made. One can recall that the loss of tensile properties caused by weathering was slightly low after

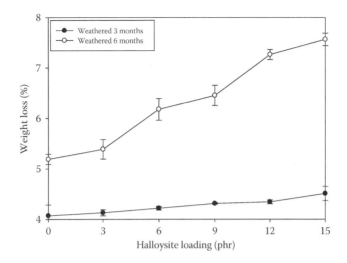

**FIGURE 8.8** Weight loss of KCF reinforced LDPE/TPSS blends with the addition of halloysite clay after exposure to outdoor natural weathering for 3 and 6 months.

3 months of exposure and was drastically reduced after 6 months. Indeed, the presence of HCs in composites has pointed out the problems related to the photooxidative stability of polymers. In this case, metal ions (iron) promote decomposition of hydroperoxide, whereas ammonium ions also decomposed and provide acidic cites on the silicate layers (Kumanayaka et al. 2010). These accelerated the oxidation reactions and breakdown of polymer chain. Therefore, the overall composite structure collapsed due to embrittlement correlated to the degradation. Apart from the aforementioned reason, the fact that individual components in composites (namely LDPE, TPSS, KCF, and HCs) experienced cyclic expansion and contraction, hydrolysis, photodegradation, and microorganisms attacks upon weathering favors the structural damages. It is believed that the deterioration in mechanical properties, as well as morphology, clearly contributed to the weight loss.

## 8.4 CONCLUDING REMARKS

The evaluations of hybrid fillers in LDPE/TPSS blends have been clearly elaborated. In the present investigation, the changes in mechanical properties along with the morphological changes after weathering have been reported. After exposure to weathering, the presence of HCs caused an abrupt decrement of strength. It was believed that HCs might accelerate the degradation reaction by acting as catalysts for the photooxidation of polymers. The oxidation and degradation of this composite was confirmed by the appearance of carbonyl groups in IR spectra. The surface deteriorations are reflected from the embrittlement of composites upon degradation.

It seems that, these types of composites have attracted considerable attention because of their processing, mechanical properties, and biodegradability advantages. However, compatibility between the components will remain as a critical issue subjected to the overall properties of the composites. Hence, further research is necessary to overcome these shortcomings. A better understanding on the issues relating to clay properties as well as their reactivity and interfacial chemistry with polymer should be taken into account. Further evaluation on their orientation and dispersion can be done using X-ray diffraction (XRD) and transmission electron microscopy (TEM). Besides, the presence of a compatibilizer has been recognized to provide better dispersion and adhesion between the clay and the polymer. Thus, the introduction of compatibilizing agent is suggested.

## REFERENCES

Ahmad Thirmizir, M.Z., Mohd Ishak, Z.A., Mat Taib, R., Sudin, R., and Leong, Y.W. (2011). Mechanical, water absorption and dimensional stability studies of kenaf bast fibre-filled poly(butylene succinate) composites. *Polymer-Plastics Technology and Engineering* 50, 339–348.

Alhuthali, A.M. and Low, I.M. (2013). Influence of halloysite nanotubes on physical and mechanical properties of cellulose fibers reinforced vinyl ester composites. *Journal of Reinforced Plastics and Composites* 32(4), 233–247.

Botta, L., Dintcheva, N.Tz., and La Mantia, F.P. (2009). The role of organoclay and matrix type in photo-oxidation of polyoefin/clay nanocomposite films. *Polymer Degradation and Stability* 94, 712–718.

Bussiere, P.O., Peyroux, J., and Chadeyron, G. (2013). Influence of functional nanoparticles on the photostability of polymer materials: Recent progress and further application. *Polymer Degradation and Stability* 98, 2411–2418.

Darie, R.N., Bercea, M., Kozlowski, M., and Spiridon, I. (2011). Evaluation of properties of LDPE/Oak wood composites exposed to artificial ageing. *Cellulose Chemistry Technology* 45(1–2), 127–135.

Dash, B.N., Rana, A.K., Mishra, H.K., Nayak, S.K., and Tripath, S.S. (2000). Novel low cost jute polyester composites. III weathering and thermal behavior. *Journal of Applied Polymer Science* 78, 1671–1679.

Eshraghi, A., Khademieslam, H., Ghasemi, I., and Talaiepoor, M. (2013). Effect of weathering on the properties of hybrid composite based on polyethylene, wood flour and nanoclay. *BioResources* 8(1), 201–210.

He, Y., Kong, W., Wang, W., Liu, T., Gong, Q., and Gao, J. (2012). Modified natural halloysite/potato starch composite films. *Carbohydrate Polymers* 87, 2706–2711.

Ismail, H. and Shaari, S.M. (2010). Curing characteristics, tensile properties and morphology of palm ash/halloysite nanotubes/ethylene-propylene-diene monomer (EPDM) hybrid composites. *Polymer Testing* 29, 872–878.

Ismail, H., Pasbakhsh, P., Ahmad Fauzi, M.N., and Abu Bakar, A. (2008). Morphological, thermal and tensile properties of halloysite nanotubes filled ethylene propylene diene monomer (EPDM) nanocomposites. *Polymer Testing* 27, 841–850.

Khan, A., Huq, T., Saha, M., Khan, R.A., and Khan, M.A. (2010). Surface modification of calcium alginate fibers with silane and methyl methacrylate monomers. *Journal of Reinforced Plastics and Composites* 29(20), 3125–3132.

Kumanayaka, T.O., Parthasarathy, R., and Jollands, M. (2010). Accelerating effect of montmorillonite on oxidative degradation of polyethylene nanocomposites. *Polymer Degradation and Stability* 95, 672–676.

Kumar, A.P., Depan, D., Tomer, N.S., and Singh, R.P. (2009). Nanoscale particles for polymer degradation and stabilization-trends and future perspectives. *Progress in Polymer Science* 34, 479–515.

Muhammad Safwan, M., Lin, O.H., and Md. Akil, H. (2013). Preparation and characterization of palm kernel shell/polypropylene biocomposites and their hybrid composites with nanosilica. *BioResources* 8(2), 1539–1550.

Mukherjee, T. and Kao, N. (2011). PLA based biopolymer reinforced with natural fibre: A review. *Journal of Polymers and the Environment* 19(3), 714–725.

Nakamura, R., Netravali, A.N., Morgan, A.B., Nyden, M.R., and Gilman, J.W. (2013). Effect of halloysite nanotubes on mechanical properties and flammability of soy protein based green composites. *Fire and Materials* 37, 75–90.

Pang, M.M., Pun, M.Y., and Ishak, Z.A.M. (2013). Natural weathering studies of bio-based thermoplastic starch from agricultural waste/polypropylene blends. *Journal of Applied Polymer Science* 129, 3237–3246.

Pasbakhsh, P., Ismail, H., Fauzi, M.N.A., and Bakar, A.A. (2009). The partial replacement of silica or calcium carbonate by halloysite nanotubes as fillers in ethylene propylene diene monomer composites. *Journal of Applied Polymer Science* 113, 3910–3919.

Pasbakhsh, P., Ismail, Ahmad Fauzi, M.N., and Abu Bakar, A. (2010). EPDM/modified halloysite nanocomposites. *Applied Clay Science* 48, 405–413.

Prachayawarakorn, J., Hommanee, L., Phosee, D., and Chairapaksatien, P. (2010). Property improvement of thermoplastic Mung Bean starch using cotton fiber and low-density polyethylene. *Starch–Stärke* 62(8), 435–443.

Rahmat, A.R., Rahman, W.A.W., Lee, T.S., and Yussuf, A.A. (2009). Approaches to improve compatibility of starch filled polymer system: A review. *Materials Science and Engineering: C* 29, 2370–2377.

# 9 Ballistic Properties of Hybrid Kenaf Composites

*R. Yahaya, S.M. Sapuan, M.R. Ishak,
Z. Leman, and M. Jawaid*

## CONTENTS

9.1 Introduction .................................................................................................... 145
9.2 Kenaf Fiber-Reinforced Composites .............................................................. 148
    9.2.1 Kenaf Fiber Hybrid Composites ........................................................ 150
    9.2.2 Kenaf–Synthetic Fiber Hybrid Composites ....................................... 150
9.3 Measurement of Ballistic Properties of Kenaf-Aramid Hybrid Composites..... 152
    9.3.1 Ballistic Properties of Kenaf–Aramid Hybrid Composites................ 154
9.4 Factors Affecting the Ballistic Properties of Kenaf Hybrid Composites ..... 157
    9.4.1 Fiber Orientation ............................................................................... 157
    9.4.2 Fiber Content..................................................................................... 157
    9.4.3 Temperature and Loading Rate......................................................... 158
    9.4.4 Fabric Architectures .......................................................................... 158
    9.4.5 Chemical Treatment .......................................................................... 159
9.5 Potential Applications of Kenaf-Aramid Hybrid Composites in Ballistics....... 160
9.6 Concluding Remarks....................................................................................... 161
Acknowledgments..................................................................................................... 162
References ................................................................................................................ 162

## 9.1 INTRODUCTION

Whether due to the awareness of environmental protection or accordance with environmental regulations, more researchers nowadays are showing interest in natural fiber composites. The use of natural fiber composites started centuries ago. For example, straw-reinforced clay was used in building construction in Egypt about 3000 years ago (Brouwer 2001). There are many factors that encourage the use of natural fiber composites; their low weight, renewable sources, low energy in production, simple process, good insulation, and acoustic properties. With a relatively short growing period, kenaf and ramie reduce the use of hardwood and preserve natural forests. Natural fiber-reinforced composites have attracted researchers due to their easy availability and environmental friendliness of their resources, processes, and final products. The depletion of petroleum-based resources, which is mainly used in synthetic fibers, also paves the way back to nature. Natural fiber-reinforced composites offer a number of advantages when compared with synthetic- and metal-based materials (Jawaid et al. 2011). Natural fiber composites offer comparable advantages, such as considerable toughness, flexibility, ease of processing, recyclability, and eco-friendliness. Research on natural fiber composites is rapidly growing, but there are some mechanicals properties that limit their application, such as in high impact components. These limitations include high moisture absorption, nonuniformity and poor mechanical properties. The major disadvantage is the polar and hydrophilic nature of lignocellulose fibers and the nonpolar characteristics of thermosetting resins (Romanzini et al. 2012). These limitations are present irrespective of whether thermoset or thermoplastic polymers are used as the matrix material (Asumani et al. 2012).

There are a variety of fibers available which can be classified into two categories; natural and synthetic. Various types of fibers are shown in Figure 9.1. Man-made fibers from chemicals are called synthetic fibers, which may be glass, carbon, aramid, boron, ceramic, etc. Synthetic fiber-reinforced composite materials, such as fiberglass, have long been used in commercial markets, and the development of high performance synthetic fibers has continued into advanced composite materials, which exhibit physical properties that are vastly superior to those of matrix materials alone. Composite materials with synthetic fibers were reintroduced in 1960s due to the scarcity of metals (and for their weight, strength, and low relative stiffness). The main purpose was to find materials with high strength and high specific stiffness. Through the evaluation of nylon with high toughness for ballistic protection, it was found that toughness is not the only criterion for ballistic protection (Bajaj 1997). High performance aramid fibers are widely used in ballistic protection. They are incorporated into a wide range of structural and impact resistance applications that require high performance and lower weight (Abu Obaid et al. 2011).

Composites that are developed using synthetic fibers, such as glass, carbon, aramid, and Kevlar®, have unique advantages over monolithic polymer materials (Begum and Islam 2013). Besides high strength and stiffness, these composites have a long fatigue life and adaptability to the intended function of the structure. Additional improvements can also be realized in the synthetic fiber composites with regards to corrosion and wear resistance, appearance, temperature-dependent behavior, environmental stability, thermal insulation, and conductivity. Although even with the known advantages, synthetic fibers do have their limitations, such as high cost, poor recyclability, and nonbiodegradable properties.

Most of the limitations both in natural and synthetic fibers can be overcome by effective hybridization of natural–natural or natural–synthetic fibers (Jawaid et al. 2012) Hybridization is a process of incorporating synthetic fibers and natural fibers in order to yield a hybrid composite with comparable strength, stiffness, strength to weight ratio, impact resistance, and other physical and mechanical properties. This type of hybridization will lead to the benefits reducing consumption of synthetic fibers; less process energy and concurrently materializing the green initiative towards a better environment. The properties of hybrid composites were studied by many

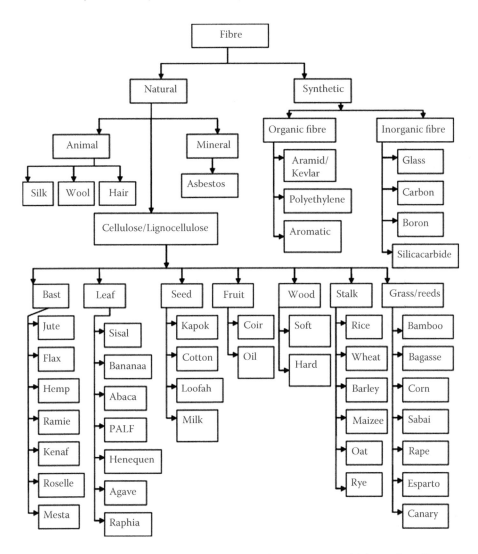

**FIGURE 9.1** Classification of natural and synthetic fibers. (From Jawaid, M. et al., *J. Compos. Mater.*, 45:2515–2522, 2011.)

researchers, such as Venkateshwaran et al. (2012), who concluded that hybrid composites offer greater resistance to water absorption, and cost saving, weight saving, and increased properties.

The paramount importance of research on natural fiber hybrid composites has led to the development of natural–synthetic fiber hybrid composites. In Chapter 9, a comprehensive overview of kenaf–synthetic fiber hybrid composites is presented. Particular emphasis is placed on the in-depth study of ballistic impact properties of kenaf–Kevlar hybrid composites. This chapter should be beneficial to researchers interested to use kenaf as an alternative or partial replacement of synthetic fibers.

## 9.2 KENAF FIBER-REINFORCED COMPOSITES

Kenaf (*Hibiscus cannabinus* L.) an annual plant which has been used since ancient times in many occasions; for example, for rope, canvas, and sacks. These plant fibers grow naturally, are renewable, and require less energy to manufacture than synthetic fibers (Joshi et al. 2004). Compared with other lignocellulose fiber crops, kenaf has advantages such as a short plantation cycle, flexibility to environmental conditions, and requiring relatively lower quantity of pesticides and herbicides (Wang and Gita 2003). Kenaf fibers are derived from the bast of the kenaf plant. They contain cellulose (44–57%), hemicellulose (22–23%), lignin (15–19%), ash (2–5%), and other elements (~6%) (Ryszard Kozłowskiy and Władyka-Przybylak 2008). In Malaysia, kenaf is the seventh largest commodity and is planted in three states with a total allocation of RM 30 milion to kenaf research for the National Kenaf and Tobacco Board (NKTB) in 2010 (Hadi et al. 2014).

Kenaf can be harvested two to three times a year; can grow to reach 3–4 m within 4–5 months; and has three layers, which are the bast, core, and pith (El-Shekeil et al. 2012). Kenaf bast fibers have been reported to have superior mechanical properties than the other parts of the plant (Aji et al. 2009). Nishino et al. (2003) claimed that the mechanical strength and thermal properties of kenaf composites are superior to other types of natural fiber polymer composites, and are suitable for high performance composites. According to Zampaloni (2007), kenaf is extremely environmentally friendly for two main reasons: kenaf accumulates carbon dioxide at a significantly high rate and kenaf absorbs nitrogen and phosphorous from the soil. Additionally, kenaf, like most other natural fibers, demonstrates low density, high specific mechanical properties, and is easily recyclable (Mohanty et al. 2000). This fiber is also commercially grown, resulting in competitive prices. There is potential for substituting synthetic fibers, such as glass and aramid, with kenaf fibers for flexural structural and nonstructural applications (Yousif et al. 2012). The performance of kenaf fiber-reinforced composites in terms of mechanical properties, such as tensile, flexural, and impact strength, have been reported (Law and Ishak 2011). Kenaf was found to be the most suitable natural fiber to be used with Kevlar in hybrid laminated composites (Yahaya et al. 2014). Tables 9.1 and 9.2 present the composition and properties of natural fibers respectively.

### TABLE 9.1
### Composition of the Natural Fibers Used (wt%)

| Component | Hemp | Flax | Kenaf |
|---|---|---|---|
| Cellulose | 65 | 56–63 | 53–57 |
| Hemicelluloses | 16 | 15–16 | 15–19 |
| Lignin, pectin | 4 | 4–6 | 5.9–9.3 |
| Wax | <1 | <1 | – |
| Proteins, minerals | 2 | 4–10 | 6 |
| Water | 12 | 12 | 7–10 |

*Source:* Cicala, G. et al., *Mater. Des. 30*:2538–2542, 2009.

## TABLE 9.2
### Properties of Natural Fibers and Kevlar

| Type of Fiber | Density (g/cm³) | Tensile Strength (MPa) | Young's Modulus (GPa) | Elongation at Break (%) | Cost (USD per kg) |
|---|---|---|---|---|---|
| Kenaf | 1.2–1.24 | 295–930 | 53 | 2.7–6.9 | 1.3 |
| Oil Palm | 0.7–1.55 | 248 | 3.2 | 2.5 | – |
| Flax | 1.4 | 800–1500 | 60–80 | 1.2–1.6 | 1.5 |
| Hemp | 1.48 | 550–900 | 70 | 1.6 | 0.6–1.8 |
| Jute | 1.46 | 400–800 | 10–30 | 1.8 | 0.35 |
| Ramie | 1.5 | 500 | 44 | 2 | 1.5–2.5 |
| Coir | 1.25 | 220 | 6 | 15–25 | 1.25 |
| Sisal | 1.33 | 600–700 | 38 | 2–3 | 0.36 |
| Cotton | 1.51 | 400 | 12 | 3–10 | 1.5–2.2 |
| Bagasse | 1.2 | 20–290 | 19.7–27.1 | 1.1 | – |
| Pineapple | 1.5 | 170–1627 | 82 | 1–3 | 0.05 |
| Banana | 1.35 | 355 | 33.8 | 5.3 | 0.10 |
| Sugar palm | 1.26 | 190.29 | 3.69 | 19.6 | – |
| Kevlar 29 | 1.44 | 3000 | 60 | 2.5–3.7 | – |

*Source:* Yahaya, R. et al., *Mater. Des. 63*:775–782, 2014.

Composites developed by various researchers, combining kenaf fibers with matrices of thermoplastic and thermoset, are summarized in Table 9.2. The selection of matrix materials depends mostly on the application of composite materials. The matrix materials play an important role in the materials of fiber-reinforced composites. In fiber-reinforced polymer matrix composites, the matrix works to transfer the load to the stiff fibers through shear stresses at the interface. Good fiber–matrix bonding will result in load absorbing capability of the composites. It also offers resistance to the composites against environment threats, such as water absorption, which will weaken the composites. Insufficient adhesion between hydrophobic polymers and hydrophilic fibers results in poor mechanical properties of the natural fiber-reinforced polymer composites (Wambua et al. 2003). Among the various thermoplastic polymers, polypropylene is perhaps one of the most widely used because of its distinct properties, such as dimensional stability, high heat distortion temperature, flame resistance, and transparency (Thakur et al. 2014).

There are numerous advantages of using natural fibers as reinforcements of the matrix, both economic and ecological, such as low density, non-abrasiveness during processing, high specific mechanical properties, and biodegradability. A more environmentally suitable option would be fully biodegradable at end-of-life by incorporating with biodegradable resins (Meredith et al. 2012). Wambua et al. (2003) reported that the mechanical properties of kenaf–polypropylene compares favorably with the corresponding properties of glass mat polypropylene composites. The properties of kenaf composites are shown in Table 9.3.

**TABLE 9.3**
**Properties of Kenaf Composites**

| Resin | Processes | Tensile Strength (MPa) | Flexural Strength (MPa) | References |
|---|---|---|---|---|
| Epoxy | Resin transfer molding | 45 | 77.4 | (Ribot et al. 2011) |
| Polyester | Vacuum infusion | 90.8 | 93.4 | (Yuhazri et al. 2011) |
| Isopthlaic polyester | Hand lay-up | 38.18 | 166.2 | (Kumar et al. 2014) |
| Polyester | Hand lay-up | 10.58 | 82.63 | (Samivel and Babu 2013) |
| Epoxy | Compression molding | 74.46 | – | (Sardar et al. 2014) |
| Epoxy | Hand lay-up | 100.56 | – | (Abdullah et al. 2012) |

*Source:* Kumar, K.P., and A.S.J. Sekaran, *J. Reinf. Plast. Compos. 33*:1879–1892, 2014.

### 9.2.1 KENAF FIBER HYBRID COMPOSITES

One way of improving the properties of composite materials is to add tough materials to the host composites. The combination of two or more of fibrous materials in common matrices leads to hybrid composites (Rao 2012). In developing hybrid composites, researchers combine fibers with high stiffness and fibers with high toughness (Muhi et al. 2009). There are several types of hybrid composites characterized as (John 2009): interply or tow-by-tow, in which tows of the two or more constituent types of fibers that are mixed in a regular or random manner; sandwich hybrids, also known as core-shell, in which one material is sandwiched between two layers of another; interply or laminated, where alternate layers of the two (or more) materials are stacked in a regular manner; intimately mixed hybrids, where the constituent fibers are made to mix as randomly as possible so that no over concentration of any one type is present in the material; and other kinds, such as those reinforced with ribs, pultruded wires, thin veils of fibers, or combinations of the above. The possible combinations of hybrid composites include artificial–artificial, natural–artificial, and natural–natural fiber types (Nunna et al. 2012).

### 9.2.2 KENAF–SYNTHETIC FIBER HYBRID COMPOSITES

The advantages of natural fibers are well-known, such as low-cost, low densities, high specific properties, non-abrasiveness, and are less harmful during handling (Joshi et al. 2004). Although natural fibers have taken a positive role in composites, they are not usually enough to be an alternative for any engineering material. They need the positive hybrid effect of synthetic additives. Hybridization with synthetic fibers and chemical modification of natural fibers has been suggested as the solution in most of the limitations (Pothan et al. 2007). Nishino et al. (2003) have found that kenaf fibers can be a good candidate for the reinforcement fibers of high performance biodegradable polymer composites, whereas Wambua et al. (2003) have investigated the mechanical properties of kenaf-reinforced polypropylene composites and found that they compare favorably with the corresponding properties of glass mat polypropylene composites. It was observed that partial replacement of artificial fibers with natural fibers led to artificial–natural fiber-based hybrid composites that show intermediate characteristics between pure natural and pure synthetic fiber-based composites (Santulli et al. 2005). Several reported works on kenaf–synthetic fibers were reported, such as those listed in Table 9.4.

Natural and synthetic fibers can be combined in the same matrix to produce hybrid composites that take full advantage of the best properties of the constituents, and thereby an optimal, superior, but economical composite can be obtained. In synthetic–natural fiber hybrid composites, most of the research aims to reduce the use of synthetic fibers (Joshi et al. 2004). The potential of natural–synthetic fiber hybridization is explained in Wambua et al. (2007). Among recent reported work on natural–synthetic fiber-based hybrid composites are pineapple–glass (Mishra et al. 2003), oil palm–glass (Khalil et al. 2007), ridge gourd–glass (Varada Rajulu and Rama Devi 2007) and jute–glass (Ahmed et al. 2007), and sisal–carbon (Noorunnisa Khanam et al. 2010).

Davoodi et al. (2010) studied hybrid kenaf–glass fiber-reinforced epoxy composite for car bumper beams. Ahmad et al. (Ahmad et al. 2011) studied polyester–kenaf fibers in unsaturated polyester composites. Liquid natural rubber (LNR) (3%) was added as a toughening agent. Kenaf fibers were treated with a sodium hydroxide solution to improve the interfacial bonding between the fiber and matrix. It was found that the addition of LNR increased impact strength and fracture toughness. Alkali fiber treatment was found to provide better impact and flexural strengths to the

### TABLE 9.4
### Reported Work in Kenaf-Synthetic Fiber Hybrid Composites

| Hybrid fiber | Matrix Polymer | References |
| --- | --- | --- |
| Kenaf/glass | Epoxy resin | (Davoodi et al. 2010) |
| Kenaf/glass | Polyester | (Akil et al. 2010) |
| Kenaf/glass | Natural rubber | (Wan Busu et al. 2010) |
| Kenaf/Fiberfrax | Phenol-formaldehyde | (Ozturk 2010) |
| Kenaf/woven glass | Polyester | (Salleh et al. 2012) |
| Kenaf/fiberglass | Polyester | (Ghani et al. 2012) |
| Kenaf/glass | Epoxy polybutylene terephthalate (PBT) | (Davoodi et al. 2012) |
| Kenaf/glass | Unsaturated polyester | (Osman et al. 2013) |
| Kenaf/glass | Unsaturated polyester | (Atiqah et al. 2014) |
| Carbon/kenaf | Epoxy | (Sapiai et al. 2014) |
| Short acetylated Kenaf bast fiber/polyaniline nanowires | Epoxy | (Izwan et al. 2014) |

composites. Dan-mallam et al. (2012) studied the mechanical properties of recycled POM–kenaf–PET hybrid composites. Kaiser et al. (2012) studied the properties of hybrid montmorillonite nanoclay (MMT) and short kenaf fiber (KF) hybrid biocomposites in a polyactic acid (PLA) matrix.

## 9.3  MEASUREMENT OF BALLISTIC PROPERTIES OF KENAF-ARAMID HYBRID COMPOSITES

The performance of ballistic-resistant composites was determined by the $V_{50}$ test, which is performed in the manner specified in the military specification MIL-STD-662F.

The value obtained in this test approximates the probabilistic velocity in which 50% of the incoming projectiles will be arrested and the other 50% will completely penetrate the panel. In the test, the panel is impacted repeatedly with a round at varying velocities until a range of partial penetrations (PP) and complete penetrations (CP) are obtained. The two or three highest velocities of PP and the corresponding two or three lowest CP velocities are taken and averaged together to determine the corresponding $V_{50}$ value. $V_{50}$ determinations were obtained using 17-grain (1.1 gram) .22 caliber fragment simulating projectiles (FSP) for this experiment at the Science and Technology Research Institute for Defence (STRIDE), Ministry of Defence. Test samples were clamped between the cover and a 10-mm-thick steel support plate. The ballistic testing setup is shown in Figure 9.2. Chisel-nosed fragment simulating

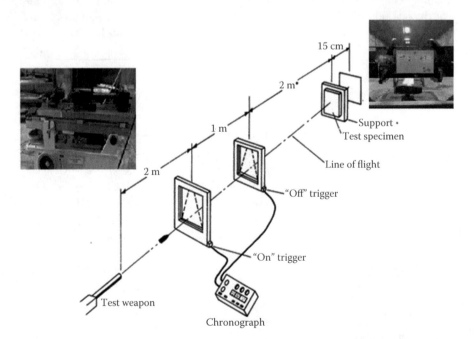

**FIGURE 9.2** Schematic test setup based on NIJ Std. 0108.01. (From National Institute of Justice, *Ballistic Resist. Prot. Mater.* 1–16, 1985.)

# Ballistic Properties of Hybrid Kenaf Composites

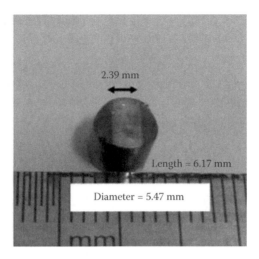

**FIGURE 9.3** Dimensional details of the FSPs used in ballistic test.

projectiles (FSPs) with weight of approximately 1.1 g and diameter of 5.40 mm, as shown in Figure 9.3, were used in this study.

The projectiles were fired from a fixed test barrel. The test samples were clamped between the cover and a 10-mm-thick steel support plate. The samples were positioned 5 m from the muzzle of the test barrel to produce impacts of 90° obliquity. A projectile velocity measurement system was used to measure the projectile at a distance of 2 m from the samples. The $V_{50}$ ballistic limit was recorded as the velocity at which an equal number of fair-impact complete penetration (target is defeated) and partial penetration (target is not defeated) were attained using the up-and-down firing method. The normal up-and-down firing procedure was used, where the propellant is manually loaded to achieve the desired velocity. The propellant filler for the next shot was adjusted until three partial and three complete penetrations were achieved within a velocity spread of no more than 40 m/s. The impact velocities ($v_s$) and residual velocities ($v_r$) were recorded, and $V_{50}$ was calculated.

The absorbed energy was calculated as the entire energy absorbed by the samples at the end of an impact event. At $V_{50}$, the impact energy is considered as totally absorbed by the composites. There are many factors that affect the two parameters, such as projectile and target geometry, strike velocity and energy, angle of impact of the projectile to the target, and the target and projectile material properties (density). In this study, the impact energy was calculated based on the muzzle velocity and the residual velocity of the projectiles. The ballistic performance of the samples was evaluated based on $V_{50}$ and energy absorption ($E_{abs}$).

The absorbed kinetic energy of the projectile can be linked by the following equations (Zhang et al. 2014):

$$E_{abs:} = E_{initial} - E_{residual} \tag{9.1}$$

where; $E_{abs}$, $E_{initial}$, and $E_{residual}$ are defined as the total kinetic energy absorbed by the armor, the kinetic energy of the projectile prior to impact, and the residual kinetic energy, respectively.

Equation 9.1 can be further derived into Equation 9.2 using the classical physics relationship that describes the kinetic energy of a moving object, whereas $m$ is the mass of the projectile, $v_s$, and $v_r$ are the projectile initial and residual velocities, respectively:

$$E_{abs} = \frac{1}{2}mv_s^2 - \frac{1}{2}mv_r^2 \tag{9.2}$$

By assuming non-deformable projectiles, the following equation was used (Zhang et al. 2014):

$$\frac{1}{2}mv_r^2 = \frac{1}{2}mv_s^2 - \frac{1}{2}mBL^2 \tag{9.3}$$

where $m$ is the mass of the projectile in kg, and $v_s$ and $v_r$ are the striking and residual velocities of the projectile in m/s, respectively.

For the calculation of ballistic limit ($V_{50}$) (Lee et al. 1994):

$$BL = \left(v_s^2 - v_r^2\right)^{\frac{1}{2}} \text{ for } v_r > 0 \tag{9.4}$$

$$V_{50} = \sqrt{\left(v_s^2 - v_r^2\right)} \tag{9.5}$$

Due to the slight variation in sample thickness, the energy absorption results presented are normalized by areal density. In this study, the specific energy absorption was calculated as (Vicente et al. 2011):

$$\text{Specific } E_{abs} = \frac{\text{absorbed energy}}{\text{areal density}} \tag{9.6}$$

The percentage of change in energy absorption ($\%E_{abs}$) was calculated as:

$$\%\text{Specific } E_{abs} = \frac{E_H - E_A}{E_A} \times 100\% \tag{9.7}$$

where $E_H$ and $E_A$ are the specific energy absorption of hybrid and non-hybrid samples, respectively.

### 9.3.1 Ballistic Properties of Kenaf–Aramid Hybrid Composites

Yahaya et al. (2014a), Yahaya et al. (2015), Yahaya et al. (2014c) and Yahaya et al. (2016) reported their research works on the mechanical and ballistic properties of

# Ballistic Properties of Hybrid Kenaf Composites

kenaf–Kevlar hybrid composites. In this study, the ballistic properties of kenaf–Kevlar hybrid composites were evaluated. The woven kenaf content varied from 5.40 to 14.99 by volume fraction by using two different arrangements: hybrid A (with one layer of woven kenaf) and B (two layers of woven kenaf). Ballistic tests were conducted using fragment simulating projectiles which were shot at measured terminal velocities ranging from 172 to 339 m/s. The results show that the hybrid composites with two layers of woven kenaf absorbed more energy when compared with samples with the 8A arrangement (eight layers of Kevlar and one layer of Kenaf) and with eight layers of Kevlar–epoxy composite. The use of two kenaf layers slightly improves the energy absorption in the hybrid composites when compared to a single layer.

Interlayer delamination was observed after the ballistic impact on hybrid composites. The effects of delamination increased the energy dissipated during ballistic impact. During perforation, delamination occurred due to larger shearing forces between the plies of fibers. The front surface was damaged due to shear plugging while the rear surface showed fiber tension. It was also observed that higher velocities were required to penetrate the thicker composites, such as samples 14A and 14B, because thicker hybrid composites provide higher resistance to the projectile penetration, thus reducing the residual velocity and absorbing more projectile kinetic energy.

Figures 9.4 and 9.5 show the failure modes of thick hybrid composites. It was observed that the penetration diameter of the front surface is smaller than that of the rear surface. This was explained as the composite failing due to penetration, followed by dishing. The additional layers of kenaf in hybrid B increased the thickness and areal density of the samples, resulting in higher energy absorption during delamination. Delamination is a significant energy absorption mechanism.

The effect of kenaf–Kevlar hybridization on the ballistic properties was evaluated based on specific energy absorption. In this study, the specific energy absorption $E_{ab.(specific)}$ was calculated as follows:

$$E_{ab.(specific)} = \frac{E_{ab.}}{Areal\, density} \quad (9.8)$$

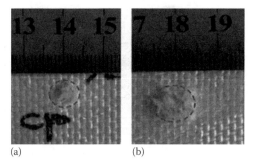

**FIGURE 9.4** Failure modes for Sample 14A: (a) front and (b) rear surfaces.

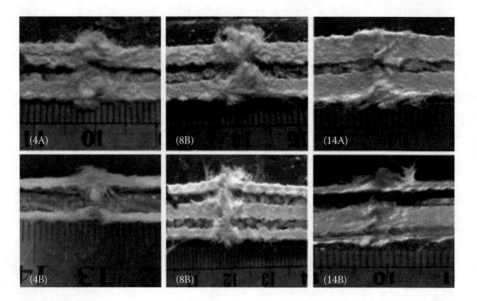

**FIGURE 9.5** Cross sectional photographs of the ballistic failure modes of the hybrid composites.

The specific energy absorption of hybrid composites is shown in Figure 9.6. The results indicate that, except the sample with four layers of Kevlar, the other hybrid type B samples absorbed higher specific energy when compared with the type A samples. The use of two kenaf layers slightly improved the specific energy absorption in sample 8B.

In this study, both woven and non-woven kenaf was evaluated. The ballistic properties of kenaf–Kevlar hybrid composites were studied and compared with those for Kevlar and kenaf–epoxy composites. The effect of hybridization was determined based on the specific ballistic energy absorption. It was found that the hybrid

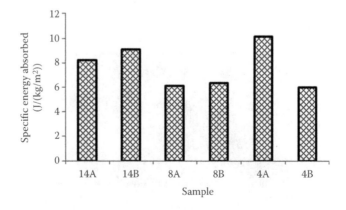

**FIGURE 9.6** Specific energy absorption of hybrid composites.

composites absorbed less energy as compared with Kevlar–epoxy. The hybrid configuration (layering sequence) also affects the energy absorption properties. Hybrid composites with kenaf at the outer layers absorbed more ballistic energy than the other hybrid configurations. The effect of woven kenaf was also studied. The woven kenaf content was varied from 5.40 to 14.99 by volume fraction by using two different arrangements: hybrid A (with one layer of woven kenaf) and B (two layers of woven kenaf). Ballistic tests were conducted using FSPs, which were shot at measured terminal velocities ranging from 172 to 339 m/s. The results show that the hybrid composites with two layers of woven kenaf absorbed more energy as compared with the samples with the 8A arrangement (eight layers of Kevlar and one layer of kenaf) and with eight layers of Kevlar–epoxy composite. The use of two kenaf layers slightly improved the energy absorption in the hybrid composites when compared to a single layer.

## 9.4 FACTORS AFFECTING THE BALLISTIC PROPERTIES OF KENAF HYBRID COMPOSITES

### 9.4.1 Fiber Orientation

Fiber orientation has significant influence on the mechanical properties of kenaf fiber-reinforced composites. Strength and stiffness of the composites are largely affected by fiber structure (Ratim et al. 2012). The highest tensile strength is achieved by woven structures while lower value is obtained in mat structure composites. The modulus elasticity of woven composites also shows a higher significant value compared to the mat and control samples. There were quite different changes of elongation at the break between woven and nonwoven fibers. However, for impact strength, the twill weave of a woven structure has shown the greatest value among all the structures. The flexural property of plain and twill weaves shows superior character among all the samples. The mechanical properties of the structural composites are largely influenced by the fiber structure and arrangement. Lee et al. (2010) investigated the properties of unidirectional kenaf fiber–polyolefin laminates. The uniaxial fiber orientation provided property enhancement of the LNPC. The randomly-oriented kenaf fibers, regardless of the fiber content of the laminates, provided an equal performance when compared to the composites made of 25% fiber glass-reinforced polyvinyl ester resin in the same laboratory processing conditions. Yahaya et al. (2014d) also studied the effects of kenaf contents and fiber orientation on physical, mechanical, and morphological properties of hybrid laminated composites for vehicle spall liners.

### 9.4.2 Fiber Content

Mechanical properties of environmentally-friendly composites made of kenaf fiber and poly-l-lactic acid (PLLA) resin was investigated by Nishino et al. (2003). They determined that optimum tensile properties and Young's modulus are dictated by the volume of reinforcing fibers used for the composites. Ishak et al. (2010) studied the mechanical properties of short kenaf bast and core fiber-reinforced unsaturated polyester composites with varying fiber weight fractions. The results showed that the optimum fiber content for achieving the highest tensile strength for both bast and core fiber composites was 20% wt. El-Shekeil et al. (2012) studied the influence of fiber content on mechanical and thermal properties of kenaf bast fiber-reinforced thermoplastic polyurethane (TPU) composites. Based on that study, composites with 30% fiber loading exhibited the best tensile strength, while the modulus increased and strain deteriorated with the increase of fiber content. The increase of fiber loading resulted in the decline in impact strength. In examining waste tire dust–kenaf fiber hybrid composites, Ismail et al. (2011) determined the increase in curing characteristics with the increment in kenaf fiber loading. Ozturk (2010) studied the effect of kenaf fiber loading on the mechanical properties of kenaf and Fiberfrax fiber-reinforced phenol-formaldehyde composites. They found that the maximum kenaf fiber content is 43%. An increase in kenaf content up to this point increases the tensile and flexural strength and hardness of the composite. However, a higher loading of kenaf fiber than 43% resulted in a drop of these properties. Ochi (2008) reported that the tensile and flexural strength of kenaf-reinforced polylactic acid (PLA) composites increased linearly, with a fiber content up to 50%. Composites of PLA containing up to 40% of kenaf fiber and up to 10% of thymol were studied to evaluate the mechanical and thermal properties.

### 9.4.3 Temperature and Loading Rate

Xue et al. (2009) studied the loading rate dependency in tensile testing of kenaf bast fiber bundles (KBFB) and kenaf fiber epoxy composite strands. The KBFB is fairly brittle and demonstrated a rate dependency in the strain rate range of $10^{-4}$–$10^{-2}$/s. The tensile strength increases gradually as the loading rate increases, while the tensile modulus almost remains the same as the loading rate increases until the loading rate reaches $10^{-2}$/s. The high temperatures (170–180°C) subjected during fiber processing and composite fabrication do not impose significant effects on the tensile properties of KBFBs if the duration is less than 1 h. The effects of post curing temperatures were reported in Yahaya et al. (2014b).

### 9.4.4 Fabric Architectures

There are factors that influence the properties of natural fiber hybrid composites. Pothan et al. (2008) studied composites of woven sisal and polyester using three different weave architectures (plain, twill, and mat) with special reference to resin viscosity, applied pressure, weave architecture, and fiber surface modification. This study provided detailed information on the effect of weaving architecture and fiber content on the hybrid composites' mechanical properties. Khan et al. (2013) studied the influence of woven structure and direction on the mechanical properties, i.e., tensile, flexural, and impact properties. It was reported that the mechanical properties of untreated woven jute composites (in warp direction) were improved compared with the nonwoven. Azrin Hani et al. (2012) studied the mechanical analysis of woven coir and kenaf natural fibers. They found that the structures used as composite reinforcements in turn produced better mechanical properties. Alavudeen et al. (2015) studied the effect of weaving patterns and random orientation on the mechanical properties of banana, kenaf, and banana–kenaf fiber-reinforced hybrid polyester composites. They found that the plain type weave architecture showed improved tensile properties when compared to twill type in all the fabricated composites. It was reported by Ratim et al. (2012) that the strength and stiffness of the polyester composite-reinforced kenaf were largely affected by fiber structure. Overall, the mechanical properties of structural composites are largely influenced by fiber structure and arrangement.

### 9.4.5 Chemical Treatment

Natural fibers can be modified either by physical or chemical means. Physical treatments change the structural and surface properties of the fiber and thereby influence their mechanical bonding to polymers. Most of the treatments done by researchers, such as Huda et al. (2008) and Reid et al. (2011), were proven to enhance the mechanical properties of composites. Two types of chemical treatments mostly used by researchers are alkali and silane treatments. In alkali treatment, kenaf fibers were immersed in sodium hydroxide solution followed by oven drying at certain temperature. In silane treatment, 5 wt.% APS (weight percentage compared to the fiber) was dissolved for hydrolysis in a mixture of water–ethanol (40:60 w/w). The pH of the solution was adjusted to 4 with acetic acid and stirred continuously during 1 h. Next, the fibers were soaked in the solution for 3 h. The fibers were then washed and kept in air for three days (Huda et al. 2008).

Nirmal et al. (2014) determined the interfacial adhesion strength of kenaf fibers using different chemical treatments in hydrochloric acid (HCl) and sodium hydroxide (NaOH) with different concentrations. Reid et al. (2011) studied the effect on the mechanical properties of kenaf fiber-reinforced polypropylene resulting from alkali–silane surface treatment. Alkali treatment can also affect the density of the composite by removing natural and artificial impurities (Farahani et al. 2012). Alkali treatment improves the surface adhesive characteristics by producing a rough surface. It also helps to improve the dispersion of fiber in the matrix, resulting in reduction in the agglomeration of the fiber.

El-Shekeil et al. (2012) studied on the influence of chemical treatment with 4% pMDI, and the second involved 2% NaOH + 4% pMDI on the tensile properties of kenaf fiber-reinforced thermoplastic polyurethane composites. The treatment of the composite with 4% pMDI did not significantly affect its tensile properties, but the treatment with 2% NaOH + 4% pMDI significantly increased the tensile properties of the composite (i.e., 30% and 42% increases in the tensile strength and modulus, respectively). Meon et al. (2012) found that the tensile properties of the treated kenaf fibers have improved significantly as compared to untreated kenaf fibers especially at the optimum level of 6% NaOH. Saiman et al. (2014) studied the positive effects of alkali treatment on the impact strength of the composite when compared to pure polyester and untreated composites. The swelling of the fibers caused the yarns to expand throughout the thickness and width of each composite. The gap between the yarns interlacing was reduced due to the expansion of yarns, which increased the covering area of the reinforced material.

Yahaya et al. (2015) reported the effect of layering sequence and chemical treatments on the mechanical properties of woven kenaf–aramid hybrid laminated composites. The woven kenaf was treated with 6% diluted NaOH before the fabrication. Layers of woven kenaf and Kevlar were hand laid-up with epoxy resin according to the specified layering sequences. The results showed that the hybrid composites with Kevlar as the outer layers possess better mechanical properties as compared to the other hybrid composites. Moreover, the tensile and flexural properties of treated (6% NaOH diluted solution) samples were better than those of nontreated kenaf hybrid composites.

## 9.5 POTENTIAL APPLICATIONS OF KENAF-ARAMID HYBRID COMPOSITES IN BALLISTICS

Studies on composites for ballistic-resistant composites have primarily focused on personal body armor, vehicle protection, and structural protection. For vehicle protection, composites are used in spall liners or secondary armor. A spall liner is a secondary armor fitted next to the conventional metallic main armor. Spall liners are designed to protect the vehicle crew in the event that the main metallic armor fails. A spall liner is a type of FRC material used for lining the interior of vehicles and/or structures to decrease the risk and effect of spall generated by the attacks on the exterior surface.

Spall liners work as protective layers next to an armor panel, which is usually composed of steel or ceramics. Spall liners work by preventing the intrusion of fragments or splinters from hard armored panels, and by decreasing the risk and the effect of the threat to a crew of armored vehicles (Erbil et al. 2011). The cause of this spall is an impact-generated compressive stress wave that passes through the armor at the speed of sound (in the struck material). When this wave strikes the rear face of the material, it is reflected as a tensile stress wave. This tensile wave can exceed the tensile strength of the material, causing a disc-shaped fracture that results in a large "scab" of material being separated from the back of the armor. Spall liners are constructed of plies of ballistic protective material such as Kevlar in the form of panels tailored to match the interior side of a protected vehicle. Figure 9.7 illustrates the spall debris cone that occurs in the vehicle's hull with and without a spall liner.

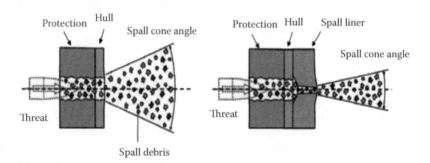

**FIGURE 9.7** Spall cone and armor protection with and without a spall liner. (From Erbil, Y. et al., A. Spall Liner: From Fiber to Protection. *6th Int. Adv. Technol. Symp.* No. May, 16–18, 2011.)

Szymczak (1995) evaluated the performance of polymeric composite materials for spall liner applications. It was found that S2 glass® and Spectra® Shield were the most efficient in mitigating behind-armor debris. It was also found that doubling the thickness significantly improved the performance.

Horsfall et al. (2007) studied the shape of the charge penetration of steel armor by using spaced metallic and composite plates of E-glass spall liner material. The authors concluded that neither the spread nor number of fragments is particularly sensitive to the spacing of the spall liner. The researchers also found that a double steel plate with a spall liner is more effective but will result in a higher areal density. Weber (2007) presented test results obtained using a kinetic energy long rod penetrator, a rolled homogenous armour (RHA) structure and homogeneous polyethylene spall liners. He concluded that in normal impacts, the mass and emission angle of the fragments increase with increasing target thickness (up to 70 mm) and decrease again for greater thicknesses.

Generally, spall liners are constructed from high performance fabrics such as aramid. Teijin Aramid (Netherlands) introduced Twaron® T765 as a new spall liner solution, which is reported to be simple and considerably less expensive to process. Twaron® yarns and fabrics have long been a key component in the production of many spall liners. Fabrics used for spall liners are usually woven structures. However, Twaron® T765 uses warp-knitting technology, making it heavier and sturdier, which in turn means it can meet high ballistic specifications with fewer layers. As another type of aramid fiber, Kevlar®, is produced by DuPont (USA). Used alone or in combination with metals or ceramics, Kevlar® spall liners help increase crew survivability in combat. Spall liners made with Kevlar do not melt or shrink when exposed to extreme heat or flames, and carbonize only at very high temperatures (approximately 900 °F in air). These liners are also extremely resistant to cuts. Kevlar® spall liners are used in M113 APC vehicles. FNSS Defence Systems (FNSS) (a Turkish defence contractor) plan to use Dyneema® BT10 tape for spall liners in 257 PARS© (Turkish for Anatolian leopard) vehicles in Malaysia.

DSM Dyneema®, the manufacturer of ultra-high molecular weight polyethylene (UHMWPE) fibers (branded as Dyneema® and world leader in life protection materials) announced a major new milestone for its lightweight, cost-effective Dyneema BT10 ballistic tape. Armored vehicles (AVs) such as the AV 8 × 8 will also be manufactured in Malaysia. These developments will offer an opportunity for researchers to evaluate hybrid composites for spall liner applications.

## 9.6 CONCLUDING REMARKS

Based on hybridization of woven kenaf–Kevlar, it is possible to obtain higher $V_{50}$ and energy absorption of aramid laminates through the appropriate design of hybrid composites by employing cheaper woven kenaf for partial substitution of aramid fibers. With the increase in kenaf content, the ballistic properties of the hybrid composites decreased. The arrangement of composite panels was also found to significantly affect the ballistic performance of hybrid composites. Furthermore, it was found that thick hybrid composites with double woven kenaf layers performed better in terms of ballistic properties. The studies of thickness and areal density of the hybrid composites indicated that the increase of parameters both increases the ballistic properties of the composites. Overall, this study provides ballistic measurement overview analysis on woven kenaf–Kevlar® hybrid composites. To ensure the practicality and feasibility of this material for actual applications, further studies are necessary to evaluate the influence of other factors on the ballistic properties of natural synthetic fiber hybrid composites.

## ACKNOWLEDGMENTS

This work was supported by the Ministry of Education Malaysia under the Research Acculturation Collaborative Effort (RACE) Grant Scheme (Project number: RACE/F1/TK2/UPNM/10). The authors would like to show their appreciation to Universiti Putra Malaysia (UPM) and the Science and Technology Research Institute for Defence (STRIDE) for supporting this research.

## REFERENCES

Abdullah, A. H., S.K. Alias, N. Jenal, K. Abdan, and A. Ali. 2012. Fatigue behavior of kenaf fibre reinforced epoxy composites. *Eng. J.* 16: 105–113.

Abu Obaid, A., J.M. Deitzel, J.W. Gillespie, and J.Q. Zheng. 2011. The effects of environmental conditioning on tensile properties of high performance aramid fibers at near-ambient temperatures. *J. Compos. Mater.* 45: 1217–1231.

Ahmed, K., S. Vijayarangan, and S. Kumar. 2007. Low velocity impact damage characterization of woven jute glass fabric reinforced isothalic polyester hybrid composites. *J. Reinf. Plast. Compos.* 26(10): 959–976.

Ahmed, K.S., S. Vijayarangan, and A.C.B. Naidu. 2007. Elastic properties, notched strength and fracture criterion in untreated woven jute–glass fabric reinforced polyester hybrid composites. *Mater. Des.* 28: 2287–2294.

Aji, I.S., S.M., Sapuan, E.S. Zainudin, and K. Abdan. 2009. Kenaf fibres as reinforcement for polymeric composites: A review. *Int. J. Mech. Mater. Eng.* 4: 239–248.

Akil, H.M., I.M. de Rosa, and C. Santulli. 2010. Sarasini, F. Flexural behaviour of pultruded jute/glass and kenaf/glass hybrid composites monitored using acoustic emission. *Mater. Sci. Eng. A.5* 27: 2942–2950.

Alavudeen, A., N. Rajini, S. Karthikeyan, M. Thiruchitrambalam, and N. Venkateshwaren. 2015. Mechanical properties of banana/kenaf fiber-reinforced hybrid polyester composites: Effect of woven fabric and random orientation. *Mater. Des.* 66: 246–257.

Asumani, O.M.L., R.G. Reid, and R. Paskaramoorthy. 2012. The effects of alkali-silane treatment on the tensile and flexural properties of short fibre non-woven kenaf reinforced polypropylene composites. *Compos. Part A Appl. Sci. Manuf.* 43: 1431–1440.

Atiqah, A., M. Maleque, M. Jawaid, and M. Iqbal. 2014. Development of kenaf-glass reinforced unsaturated polyester hybrid composite for structural applications. *Compos. Part B Eng.* 56: 68–73.

Azrin Hani, A.R., C.T. Seang, R. Ahmad, and J.M. Mariatti. 2012. Impact and Flexural Properties of Imbalance Plain Woven Coir and Kenaf Composite. *Appl. Mech. Mater.* 271: 81–85.

Bajaj, P. 1997. Ballistic protective clothing: An overview. *Indian J. Fibre Text. Res.* 22: 274–291.

Begum, K. and M. Islam, 2013. Natural Fiber as a substitute to Synthetic Fiber in Polymer Composites: A Review. *Res. J. Eng. Sci.* 2: 46–53.

Brouwer, W.D. 2001. Natural fibre composites, saving weight and cost with renewable materials. *Thirteenth International Conference on Composite Materials*; Beijing, China, 1414.

Cicala, G., G. Cristaldi, G. Recca, G. Ziegmannb, and A. El-Sabbagh. 2009. Dickert, M. Properties and performances of various hybrid glass/natural fibre composites for curved pipes. *Mater. Des.* 30: 2538–2542.

Dan-mallam, Y., M.Z. Abdullah, and S.M. Megat Yusoff. 2012. Predicting the tensile properties of woven kenaf/polyethylene terephthalate (PET) fiber reinforced polyoxymethylene (POM) hybrid laminate composite. *J. Apllied Polym. Sci.* 2: 6–13.

Davoodi, M.M., S.M. Sapuan, D. Ahmad, A. Aidy, A. Khalina, and M. Jonoobi. 2012. Effect of polybutylene terephthalate (PBT) on impact property improvement of hybrid kenaf/glass epoxy composite. *Mater. Lett.* 67: 5–7.

Davoodi, M.M., S.M. Sapuan, D. Ahmad, A. Ali, A. Khalina, And M. Jonoobi. 2010. Mechanical properties of hybrid kenaf/glass reinforced epoxy composite for passenger car bumper beam. *Mater. Des.* 31: 4927–4932.

El-Shekeil, Y.A., S. M. Sapuan, K. Abdan, and E.S. Zainudin. 2012. Influence of fiber content on the mechanical and thermal properties of Kenaf fiber reinforced thermoplastic polyurethane composites. *Mater. Des.* 40: 299–303.

El-Shekeil, Y.A., S.M. Sapuan, A. Khalina, E.S. Zainudin, and O.M. Al-Shuja'a. 2012. Influence of chemical treatment on the tensile properties of kenaf fiber reinforced thermoplastic polyurethane composite. *Express Polym. Lett.* 6: 1032–1040.

Erbil, Y., A.K. Ekşi, and D.A. Bircan. 2011. A Spall Liner: From Fiber to Protection. *6th Int. Adv. Technol. Symp.* No. May, 16–18.

Farahani, G.N., I. Ahmad, and Z. Mosadeghzad. 2012. Effect of Fiber Content, Fiber Length and Alkali Treatment on Properties of Kenaf Fiber/UPR Composites Based on Recycled PET Wastes. *Polym. Plast. Technol. Eng.* 51: 634–639.

Ghani, M.A.A., Z. Salleh, K.M. Hyie, M.N. Berhan, T.M.D. Taib, and M.A.I Bakri. 2012. Mechanical properties of kenaf/fiberglass polyester hybrid composite. *Procedia Eng.* 41: 1654–1659.

Hadi, M., A. Basri, A. Abdu, N. Junejo, and H.A. Hamid. 2014. Journey of kenaf in Malaysia: A Review. *Sci. Res. Essay* 9: 458–470.

Horsfall, I., E. Petrou, and S.M. Champion. 2007. Shaped charged attack of spaced and composite armour. In *23rd International Symposium on Ballistics*; Tarragona, Spain, 1281–1288.

Huda, M.S., L.T. Drzal, A.K. Mohanty, and M. Misra. 2008. Effect of fiber surface-treatments on the properties of laminated biocomposites from poly(lactic acid) (PLA) and kenaf fibers. *Compos. Sci. Technol.* 68: 424–432.

Ishak, M.R., Z. Leman, S.M. Sapuan, A.M.M. Edeerozey, and I.S. Othman. 2010. Mechanical properties of kenaf bast and core fibre reinforced unsaturated polyester composites. *IOP Conf. Ser. Mater. Sci. Eng.* 11: 12006.

Ismail, H., N.F. Omar, and N. Othman. 2011. The effect of kenaf fibre loading on curing characteristics and mechanical properties of waste tyre dust/kenaf fibre hybrid filler filled natural rubber compounds. *BioResources* 6: 3742–3756.

Izwan, S., A. Razak, W. Aizan, and W. Abdul. 2014. Hybrid composites of short acetylated kenaf bast fiber and conducting polyaniline nanowires in epoxy resin. *J. Compos. Mater.* 48: 667–676.

Jawaid, M. and H.P.S.A. Khalil. 2011. Cellulosic/synthetic fibre reinforced polymer hybrid composites: A review. *Carbohydr. Polym.* 86: 1–18.

Jawaid, M., H.P.S. Abdul Khalil, and Azman Hassan. 2012. Bi-layer hybrid biocomposites: chemical resistant and physical properties. *BioResources* 7: 2344–2355.

Jawaid, M., H.P.S.A. Khalil, and A.A. Bakar. 2011. Hybrid composites of oil palm empty fruit bunches/woven jute fiber: Chemical resistance, physical, and impact properties. *J. Compos. Mater.* 45: 2515–2522.

John M.J., A.R., and T. S. Hybrid composites. 2009. *Natural Fibre Reinforced Polymer Composite: Macro to Nanoscale.* Pothan, S.T. and L.A., Ed. Old City Pub Inc: 315–328.

Joshi, S., L. Drzal, A. Mohanty, and S. Arora. 2004. Are natural fiber composites environmentally superior to glass fiber reinforced composites? *Compos. Part A Appl. Sci. Manuf.* 35: 371–376.

Kaiser, M. R., H.B. Anuar, N.B. Samat, and S.B.A. Razak. 2012. Effect of processing routes on the mechanical, thermal and morphological properties of PLA-based hybrid biocomposite. *Iran. Polym. J.* 22: 123–131.

Khalil, H.P.S.A., S. Hanida, C.W. Kang, and N.A.N. Fuaad. 2007. Agro-hybrid composite: The effects on mechanical and physical properties of oil palm fiber (EFB)/glass hybrid reinforced polyester composites. *J. Reinf. Plast. Compos.* 26: 203–218.

Khan, G.M.A., M. Terano, M.A. Gafur, and M.S. Alam. 2013. Studies on the mechanical properties of woven jute fabric reinforced poly(l-lactic acid) composites. *J. King Saud Univ. Eng. Sci.* 28: 69–74.

Kumar, K.K., P.R. Babu, and K.R. Narender. 2014. Evaluation of Flexural and Tensile Properties of Short Kenaf Fiber Reinforced Green Composites. *Int. J. Adv. Mech. Eng.* 4: 371–380.

Kumar, K.P., and A.S.J. Sekaran. 2014. Some natural fibers used in polymer composites and their extraction processes: A review. *J. Reinf. Plast. Compos.* 33: 1879–1892.

Law, T.Z.A. and Z.A.M. Ishak. 2011. Water absorption and dimensional stability of short kenaf fiber-filled polypropylene composites treated with maleated polypropylene. *J. Appl. Polym. Sci.* 120: 563–572.

Lee, B. L., J.W. Song, and J.E. Ward. 1994. Failure of Spectra(R) Polyethylene Fiber-Reinforced Composites under Ballistic Impact Loading. *J. Compos. Mater.* 28: 1202–1226.

Lee, S., S.Q. Shi, L.H. Groom, and Y. Xue. 2010. Properties of Unidirectional Kenaf Fiber–Polyolefin Laminates. *Polym. Compos.* 31: 1067–1074.

Meon, M.S., M.F. Othman, H. Husain, M.F. Remeli, and M.S.M. Syawal. 2012. Improving Tensile Properties of Kenaf Fibers Treated with Sodium Hydroxide. *Procedia Eng.* 41: 1587–1592.

Meredith, J., R. Ebsworth, S.R. Coles, B.M. Wood, and K. Kirwan. 2012. Natural fibre composite energy absorption structures. *Compos. Sci. Technol.* 72: 211–217.

Mishra, S., A.K. Mohanty, L. Drzal, M. Misra, S. Parija, S. Nayak, and S. Tripathy. 2003. Studies on mechanical performance of biofibre/glass reinforced polyester hybrid composites. *Compos. Sci. Technol.* 63(10): 1377–1385.

Mohanty, A., K., M. Misra, and G. Hinrichsen. 2000. Biofibres, biodegradable polymers and biocomposites: An overview. *Macromol. Mater. Eng.* 276–277: 1–24.

Muhi, R. J., F. Najim, and M.F.S.F. de Moura. 2009. The effect of hybridization on the GFRP behavior under high velocity impact. *Compos. Part B Eng.* 40: 798–803.

National Institute of Justice. Ballistic Resistant Protective Materials. 1985. NIJ Standard 0108.01. *Ballistic Resist. Prot. Mater.* 1–16.

Nirmal, U., S.T.W. Lau, and J. Hashim. 2014. Interfacial Adhesion Characteristics of Kenaf Fibres Subjected to Different Polymer Matrices and Fibre Treatments. *J. Compos.* 2014: 1–12.

Nishino, T., K. Hirao, Kotera, M., K. Nakamae, and H. Inagaki. 2003. Kenaf reinforced biodegradable composite. *Compos. Sci. Technol.* 63: 1281–1286.

Noorunnisa Khanam, P., H.P.S. Abdul Khalil, M. Jawaid, G. Ramachandra Reddy, C. Surya Narayana, S. Venkata Naidu. 2010. Sisal/Carbon Fibre Reinforced Hybrid Composites: Tensile, Flexural and Chemical Resistance Properties. *J. Polym. Environ.* 18: 727–733.

Ahmad, S.H., R. Rasid, N.N. Bonnia, I. Zainol, A.A. Mamun, A.K. Bledzki, and M.D.H Beg. 2011. Polyester-kenaf composites: effects of alkali fiber treatment and toughening of matrix using liquid natural rubber. *J. Compos. Mater.* 45: 203–217.

Nunna, S., P. Chandra, S. Shrivastava, and A. Jalan. 2012. A review on mechanical behavior of natural fiber based hybrid composites. *J. Reinf. Plast. Compos.* 31: 759–769.

Ochi, S. 2008. Mechanical properties of kenaf fibers and kenaf/PLA composites. *Mech. Mater.* 40: 446–452.

Osman, M.R., H.M. Akil, Z.A. Mohd Ishak. 2015. Effect of hybridization on the water absorption behaviour of pultruded kenaf fibre-reinforced polyester composites. *Compos. Interfaces.* 20: 517–528.

Ozturk, S. 2010. Effect of fiber loading on the mechanical properties of kenaf and Fiberfrax fiber-reinforced phenol-formaldehyde composites. *J. Compos. Mater.*, 44: 2265–2288.

Pothan, L.A., B.M. Cherian, B. Anandakutty, and S. Thomas. 2007. Effect of Layering Pattern on the Water Absorption Behavior of Banana Glass Hybrid Composites. *J. Appl. Polym. Sci.* 105: 540–2548.

Pothan, L.A., Y.W. Mai, S. Thomas, and R.K.Y. Li. 2008. Tensile and flexural behavior of sisal fabric/polyester textile composites prepared by resin transfer molding technique. *J. Reinf. Plast. Compos.* 27: 1847–1866.

Rao, A.N.H. 2012. A review on recent applications and future prospectus of hybrid composites. *Int. J. Soft Comput. Eng.* 1: 352–355.

Ratim, S., N.N. Bonnia, and N. S. Surip. 2012. The effect of woven and non-woven fiber structure on mechanical properties polyester composite reinforced kenaf. In *AIP Conference Proceedings.* 1455: 131–135.

Reid, R.G., O.M.L. Asumani, and R. Paskaramoorthy. 2011. The effect on the mechanical properties of kenaf fibre reinforced polypropylene resulting from alkali-silane surface treatment. Ferreira A.J.M., Ed. 16th *International conference on composite structures ICCS16*, Porto, Portugal, 24–25.

Ribot, N.M.H., Z. Ahmad, and N.K. Mustaffa. 2011. Mechanical Propertise of Kenaf Fiber Composite Using Co-Cured in-Line Fiber Joint. *Int. J. Eng. Sci. Technol.* 3: 3526–3534.

Romanzini, D.H.L., Ornaghi Junior, S.C. Amico., and A.J. Zattera,. 2012 Preparation and characterization of ramie-glass fiber reinforced polymer matrix hybrid composites. *Mater. Res.* 15: 415–420.

Ryszard Kozłowskiy and M. Władyka-Przybylak. 2008. Flammability and fire resistance of composites reinforced by natural fiber. *Polym. Adv. Technol.* 19: 446–453.

Saiman, M.P., M.S. Wahab, and M.U. Wahit. 2014. The Effect of Fabric Weave on Tensile Strength of Woven Kenaf Reinforced Unsaturated Polyester Composite. *Appl. Mech. Mater.* 564: 52–56.

Salleh, Z., Y.M. Taib, K.M. Hyie, M. Mihat, M.N. Berhan, and M.A.A Ghani. 2012. Fracture toughness investigation on long kenaf/woven glass hybrid composite due to water absorption effect. *Procedia Eng.* 41: 1667–1673.

Samivel, P. and A.R. Babu. 2013. Mechanical behavior of stacking sequence in kenaf and banana fiber reinforced-polyester laminate. *Int. J. Mech. Eng. Rob. Res.* 2: 348–360.

Santulli, C., M. Janssen, and G. Jeronnimidis. 2005. Partial replacement of E-glass fibers with flax fibers in composites and effect on falling weight impact performance. *J. Mater. Sci.* 40: 3581–3585.

Sapiai, N., A. Jumahat, and R.N. Hakim. 2014.Tensile and Compressive Properties of Hybrid Carbon Fiber/Kenaf Polymer Composite. *Adv. Environ. Biol.* 8: 2655–2661.

Sardar, K., K. Veeresh, and M. Gowda. 2014. Characterization and Investigation of Tensile Test on Kenaf Fiber Reinforced Polyester Composite Material. *IJRDET.* 2: 104–112.

Szymczak, M. 1995. *A Preliminary Evaluation of Selected Polymeric Composites as Behind-Armour Debris Mitigators.* Valcartier, Quebec.

Thakur, V.K., M.K. Thakur, and R.K. Gupta. 2014. Review: Raw Natural Fiber–Based Polymer Composites. *Int. J. Polym. Anal. Charact.* 19: 256–271.

Varada Rajulu, A. and R. Rama Devi. 2007, Tensile properties of ridge gourd/phenolic composites and glass/ridge gourd/phenolic hybrid composites. *J. Reinf. Plast. Compos.* 26: 629–638.

Venkateshwaran, N., A. Elayaperumal, and G.K. Sathiya. 2012. Prediction of tensile properties of hybrid-natural fiber composites. *Compos. Part B Eng.* 43: 793–796.

Vicente, J.F., A. Jerusalem, Y. Zhang, M. Dao, J. Lu, and F. Galvez. 2011. Ballistic Testing of Nanocrystalline Hybrid Composites. In *Proceedings of the 26th International Symposium on Ballistics (ISB)*. Baker, E., Templeto, D., Eds. DEStech Publications, Inc.: Miami, 1906–1917.

Wambua, P., B. Vangrimde, S. Lomov, and I. Verpoest. 2007. The response of natural fibre composites to ballistic impact by fragment simulating projectiles. *Compos. Struct.* 77: 232–240.

Wambua, P., J. Ivens, and I. Verpoest. 2003. Natural fibres: Can they replace glass in fibre reinforced plastics ? *Compos. Sci. Technol.* 63:1259–1264.

Wan Busu, W.N.H. Anuar, S.H. Ahmad, R. Rasid, and N.A. Jamal. 2010. The Mechanical and Physical Properties of Thermoplastic Natural Rubber Hybrid Composites Reinforced with *Hibiscus cannabinus*, Long and Short Glass Fiber. *Polym. Plast. Technol. Eng.* 49: 1315–1322.

Wang, J. and N. Gita. 2003, One-Step Processing and Bleaching of Mechanically Separated Kenaf Fibers: Effects on Physical and Chemical Properties. *Text. Res. J.* 73: 339–344.

Weber, K. 2007. Behind armor debris distribution after KE rod perforation of RHA plates for distinct overmatch conditions. In *23rd International Symposium on Ballistics*; Tarragona, Spain, 2007; 1115–1122.

Xue, Y., Y. Du, S. Elder, K. Wang, and J. Zhang. 2009. Temperature and loading rate effects on tensile properties of kenaf bast fiber bundles and composites. *Compos. Part B Eng.* 40: 189–196.

Yahaya, R., S. M. Sapuan, Z. Leman, and E.S. Zainudin. 2014. Selection of Natural Fibre for Hybrid Laminated Composites Vehicle Spall Liners Using Analytical Hierarchy Process (AHP). *Appl. Mech. Mater.* 564: 400–405.

Yahaya, R., S.M. Sapuan, M. Jawaid, Z. Leman, and E.S. Zainudin. 2014. Mechanical performance of woven kenaf-Kevlar hybrid composites. *J. Reinf. Plast. Compos.* 33: 2242–2254.

Yahaya, R., S.M. Sapuan, M. Jawaid, Z. Leman, and E.S. Zainudin. 2015. Effect of layering sequence and chemical treatment on the mechanical properties of woven kenaf–aramid hybrid laminated composites. *Mater. Des.* 67: 173–179.

Yahaya, R., S.M. Sapuan, M. Jawaid, Z. Leman, and E.S. Zainudin. 2014. Quasi-Static Penetration and Ballistic Properties of Kenaf-Aramid Hybrid Composites. *Mater. Des.* 63: 775–782.

Yahaya, R., S.M. Sapuan, M. Jawaid, Z. Leman, and E.S. Zainudin. 2016. Measurement of ballistic impact properties of woven kenaf-aramid hybrid composites. *Meas. J. Int. Meas. Confed.* 77: 335–343.

Yahaya, R., S.M. Sapuan, M. Jawaid, Z. Leman, and E.S. Zainudin. 2014. Effects of kenaf contents and fiber orientation on physical, mechanical, and morphological properties of hybrid laminated composites for vehicle spall liners. *Polym. Compos.* 36: 1469–1476.

Yahaya, R., S.M. Sapuan, M. Jawaid, Z. Leman, and E.S. Zainudin. 2014. Effect of post curing, fibre content and resin-hardener mixing ratio on the properties of kenaf-aramid hybrid composites. *Appl. Mech. Mater.* 548: 7–11.

Yousif, B.F., A. Shalwan, C.W. Chin, and K.C. Ming. 2012. Flexural properties of treated and untreated kenaf/epoxy composites. *Mater. Des.* 40: 378–385.

Yuhazri, M.H. Sihombing, A.R. Jeefferie, and K. Rassiah. 2011. Mechanical properties of kenaf/polyester composites. *Int. J. Eng. Technol.* 11: 127–131.

Zampaloni, M. 2007. Kenaf natural fiber reinforced polypropylene composites: A discussion on manufacturing problems and solutions. 38: 1569–1580.

Zhang, D., Y. Sun, L. Chen, S. Zhang, and N. Pan. 2014. Influence of fabric structure and thickness on the ballistic impact behavior of Ultrahigh molecular weight polyethylene composite laminate. *Mater. Des.* 54: 315–322.

# 10 Cellulose-Based Composites from Kenaf Fibers

*J. Sahari, M.A. Maleque, and M.L. Sanyang*

## CONTENTS

10.1 Introduction: Kenaf Fibers .................................................................... 169
10.2 Extraction of Kenaf Fibers ..................................................................... 170
10.3 Extraction of Cellulose from Kenaf Fibers ............................................. 173
10.4 Cellulose Kenaf Fiber-Reinforced Thermoplastic Composites ................ 175
10.5 Cellulose Kenaf Fiber-Reinforced Thermoset Composites ..................... 176
10.6 Cellulose Kenaf Fiber-Reinforced Biopolymers ..................................... 178
10.7 Cellulose Kenaf Fiber-Treated Composites ............................................ 180
References ........................................................................................................ 181

## 10.1 INTRODUCTION: KENAF FIBERS

The word "kenaf" comes from the Persian dialect in the late nineteenth century (Mahjoub et al. 2014). Kenaf plants have become the subject of attention in Malaysia, particularly by the government, as kenaf plants are believed to be a potential substitute for tobacco plants in the future. The nationwide plan is to enlarge the development of kenaf in a variety of applications comprising of automotive components, furniture, food packaging, and sports and leisure. Kenaf fibers are also used as a reinforcement for plastic and synthetic products, cosmetics, organic fillers, and medicine (Salleh et al. 2014).

Kenaf is a globally preferred material, which was initially cultivated in the United States in 1940s. Mossello et al. (2010) reported that 80% of the world's kenaf produced today is from India, China, and Thailand. A few states with warm climates, like Georgia, Florida, Mississippi, Texas, and New Mexico, also cultivate the kenaf plant. Malaysia is in the process of decreasing the global consumption of nondegradable plastics, particularly for domestic appliances (Anuar and Zuraida 2011; Mohd Edeerozey et al. 2007). Since the 1960s, an increasing demand for kenaf has started to develop, mainly for its capability to use as an industrial fiber crop for the production of newsprint as well as the other pulp and paper products, which lead to collaborations between the research and development, economic, and market research sectors (Abdul Khalil et al. 2010).

The Kenaf plant belongs to the family of Malvaceae, also known by a scientific name as *Hibiscus cannabinus* L. with a high rate of growth, since the height of the plant rise from 4–6 m in only about 4–5 months. Kenaf has a fibrous stalk which is hardy, strong, and tough and resistant to insect damage, therefore eliminating the need for pesticides (Elsaid et al. 2011). Bella et al. (2014) stated that the growth of kenaf plant does not need pest control as it absorbs chemical and heavy metals from the soil. In addition, kenaf plants have an extensive range in the adaptation of climates and soils compared to other fibrous plant used for industrial manufacture. The dry weight produced by the kenaf plant is 6000–10,000 kg/ha a year, the new range may reach 30,000 kg/ha a year (Bella et al. 2014).

Kenaf has natural features which make it an eco-friendly material. The plant has a high rate absorption of carbon dioxide. About 1.5 tons of carbon dioxide are absorbed during the manufacture of one ton of dry kenaf, and thus kenaf is highest absorber of carbon dioxide than any known plant (Mohanty et al. 2005). In addition, kenaf also absorbs high amount of nitrogen and phosphorus in the soil, which means that kenaf can avoid water pollution when it is planted close to rivers or streams. It can absorb nitrogen in the rate of 0.81 g/m$^2$/day and 0.11 g/m$^2$/day of phosphorus, which shows a higher rate than most other trees (Baillie 2005).

In comparisons of production, anatomy, stem processing, fiber quality, gain, and pricing, the kenaf plant clearly has manifold benefits versus jute, hemp, and flax as a source of fiber (Karimi et al. 2014). One of the common natural fibers used in reinforcement in polymer matrix composites is kenaf, and it is a necessary resource of cellulose composites and the other industrial applications. The yearly kenaf crop is harvested during the warm season, which is cultivated in temperate and tropical areas. It is systematically linked to cotton, okra, and hibiscus.

Since kenaf is a fibrous plant, it is comprised of inner core fibers (75–60%), which create low quality of pulp, and the outer bast of the stem (25–40%) produces a high quality of pulp. The harvested time for kenaf plant is 2–3 times a year. When the height of the kenaf plant reaches 2.7–3.6 m, the fibers are ready to be extracted from the stalks. The requirement of kenaf in attaining the suitable size for practical uses is only less than 6 months. The cell wall of kenaf in common with wood comprises of rigid fillers, name as cellulose fibers, inserted into soft matrix substances of hemicellulose and lignin (Lam et al. 2003).

Kenaf can be efficiently used as an alternative to wood in production of pulp and paper (Tahir et al. 2011). The paper produced by kenaf has some benefits over wood paper since only a small amount of chemicals are utilized during the pulping process which lowers the amount of environmental pollution. Additionally, the paper of kenaf produced is of a higher quality than wood paper. Kenaf paper is whiter, tougher, more stable, more resistant to yellowing, and has a finer ink adherence (Tahir et al. 2011).

## 10.2 EXTRACTION OF KENAF FIBERS

Kenaf fibers are advantageous from both an economic and environmental standpoint (Saba et al. 2015). They are becoming increasingly useful in Malaysia as a natural resource that may provide an eco-friendly alternative for the manufacture automotive, food packaging, furniture, and sports industries (Anuar and Zuraida 2011). Tahir et al. (2011) stated that the part of the kenaf plant which has a high quality of kenaf fibers is the stalk, which can be harvested during the beginning of the flowering season. A low quality of fiber yielded when the fibers are extracted after the flowering season.

The kenaf stem is cylindrical in structure, and is comprised of two types of fibers: the internal fibers (core) and the outer fibers (bast), as shown in Figure 10.1. The scanning electron micrograph (SEM) of kenaf core and bast fibers is shown in Figure 10.2. These two types of fibers have different applications, the internal core fibers are very soft and hollow which are suitable to be used as organic filler in plastic, while the outer bast fibers are suitable to be used in blending with plastics in the textile industry, as well as for fiberglass technology, since the bast fibers are hard.

Kenaf bast strands are utilized as support or filler for polymer composite materials, where they contend in the form of properties and costs with other fibers like flax, hemp, and jute (Summerscales et al. 2010). Automotive production shows the improvement market for fiber-based composites. Natural fiber composites are utilized as a part of car applications due to their lightweight and end-of-life properties (reusing) (Lips and Dam 2013). The application of these natural fiber-based composites for nonwoven fiber mats is used for internal automotive parts, for example, headliners, divider boards, trunk liners, and hoods. Kenaf and jute fibers are commonly used in the automotive industry, which are compacted with thermoforming polypropylene (PP) fibers, however, the measure of kenaf is unspecified. In the most recent decade, many studies on composites with distinctive grids were distributed. Other than polypropylene as support, kenaf fibers were combined with polyester (UPE), polyurethane (UR) (El-Shekeil et al. 2012), and epoxy (Yousif et al. 2012).

The pith of kenaf fibers can be compared to the fibers found in hardwood (Mohd Edeerozey et al. 2007). Kenaf bast fibers have superior mechanical properties to other parts of the plant (Aji et al. 2009). Kenaf fibers are comprised of a 56–64% cellulose content, lignin content is around 5.9 and 19% and hemicellulose content

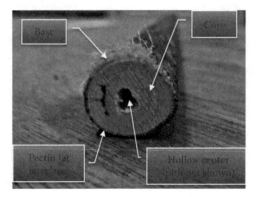

**FIGURE 10.1** Cross section of kenaf stalk. (From Sheldon, A., Preliminary Evaluation of Kenaf as a Structural Material. A Thesis Presented to the Graduate School of Clemson University In Partial Fulfillment of the Requirements for the Degree Master of Science Civil Engineering, 1–57, 2014.)

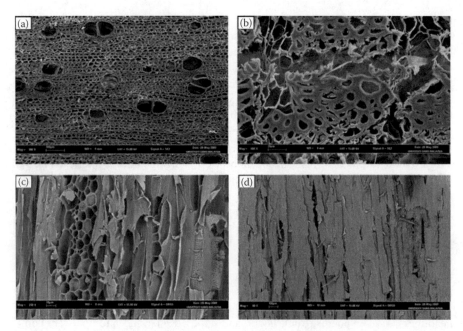

**FIGURE 10.2** Scanning electron micrograph of kenaf core and bast fibers: (a) transverse section of core fibers; (b) transverse section of bast fibers; (c) longitudinal section of core fibers; (d) longitudinal section of bast fibers. (From Abdul Khalil, H.P.S. et al., *Industrial Crops and Products*, 31, 113–121, 2010.)

around 15 and 23% as well as little measures of extractives and ash, respectively (Mazuki et al. 2011; Davoodi et al. 2010; Zainuddin et al. 2013). Table 10.1 shows the mechanical and chemical composition of kenaf reported by Mohanty et al. (2005). Since the content of cellulosic of kenaf fiber is relatively high, it is utilized as a basis of paper, grass mats, oil-absorbent materials, animal bedding, and textiles (Keshk et al. 2006).

A few diverse introductory retting methods were accounted by Holbery and Houston (2006) in order to extract the kenaf fibers, which include several methods, the first one is dew retting in the field when the plants are gathered. Downpour, dew, or watering systems are utilized to keep them wet. This technique takes around 5 weeks and the fiber produced is low quality. Water retting is a prominent system

**TABLE 10.1**
**Mechanical Properties and Chemical Composition of Kenaf**

| Tensile Strength (MPa) | Tensile Modulus (GPa) | Tensile Elongation (%) | Extractive (%) | Holocellulose (%) | α-Celluloses (%) | Lignin (%) | Ash (%) |
|---|---|---|---|---|---|---|---|
| 930 | 53 | 1.6 | 5.5 | 86.8 | 55 | 14.7 | 5.4 |

*Source:* Mohanty, A.K. et al., *Natural Fibers, Biopolymers, and Biocomposites*, CRC Press, Broken Sound Parkway, 2005.

through which microbes separate the pectin. This method takes around 10 days and created a great quality of fiber. The warm-water retting technique starts with the warm water submersion of the kenaf that goes on for at least 3 days and fiber will be uniform and clean. The most common process used to prepare fibers for textile, paper, and fiberboard applications are the mechanical processes which produce short or twisted fibers. In chemical retting, a few chemicals are utilized to dissolve the pectin and separate the parts. This step is a brisk and costly process and creates high quality fibers.

The methods of extracting the kenaf fibers include microbial techniques and enzymatic and chemical degumming processes. The cellulose fiber structure must not be infected or damage during the elimination of noncellulosic polysaccharides, since the quality of fiber produced depends on the method of the extraction of fiber. Good quality fiber bundles are formed using the traditional method of fiber extraction, but traditional water retting is conducted in streams and tanks because the surface of the water is polluted. In order to gain low-cost separation, the whole stems must be pulverized followed by the separation of the bast and core fibers. The strength of the fiber bundles is affected during this green decortication process because of the rough mechanical process (Lips and Dam 2013).

## 10.3  EXTRACTION OF CELLULOSE FROM KENAF FIBERS

Mahjouba et al. (2014) stated that all natural fibers are cellulosic in nature and are made up of cellulose, hemicellulose, lignin, and pectin. The main components are cellulose and lignin. In addition, lignin and hemicellulose serve as the glue between cellulose and the matrix (Shin et al. 2012). The quantity of the cellulose of natural fibers differs based on the species and age of the plants. Cellulose is a type of semi-crystalline polysaccharide hydrophilic constituent which presents both amorphous and crystalline phases and consists of linear chain of anhydroglucose units which have hyroxyl groups, where the ring of $\delta$-glucopyranose are linked with $\beta$-(1,4)-glycosidic bonds, as shown in Figure 10.3. (Akil et al. 2011; Siro and Plackett 2010)

The cellulose content of natural fibers affect their mechanical properties, such as fiber length, fiber stacking (i.e., volume portion of fibers), fiber aspect ratio, fiber orientation, or interfacial bond between fiber and matrix (Rassiah and Ahmad 2013). Cellulose is inexpensive, biodegradable, and edible. Since cellulose is infusibile and

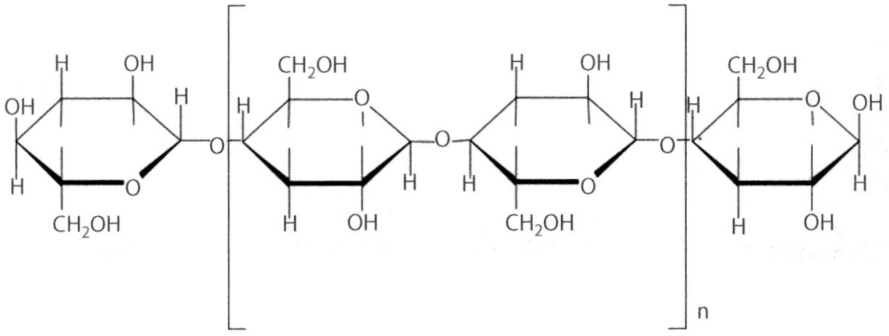

**FIGURE 10.3** Chemical structure of cellulose. (From Akil, H.M. et al., *Materials and Design*, 32 (8–9), 4107–4121. 2011.)

insoluble, it is frequently transformed into derivatives in order to make it easily procesable such as methyl cellulose, cellulose acetate, and so on (Weber et al. 2002). It is most widely used as a starting material for the production the other valuable products like rayon, microcrystalline cellulose (MCC), carboxymethyl cellulose (CMC), and cellulose derivatives such as α-cellulose (Sixta, 2006). Commonly, α-cellulose improves the stiffness and the strength of different polymeric matrices, and has been derived from natural fibers such as kenaf (Khalid et al. 2008) and oil-palm (Tajeddin et al. 2009).

Another method for extracting cellulose from the kenaf bast fibers is by using an electron beam irradiation (EBI) treatment (Shin et al. 2012). EBI has been used to reduce cell wall constituents or corrupt and delignify cellulose-based fibers without giving negative effects. The fundamental point of interest is that EBI incorporates the capacity to advance changes in reproduction and quantity without using chemical reagents or the requirement of extraordinary hardware to control temperature, environment, and additives (Shin et al. 2012). This technique of separation based on the hot–warm treatment after EBI is followed by a two-step bleaching process. Bleached cellulose fibers in Figure 10.3b are obtained by alkali treatment, while 10.3c, 10.3d, 10.3e, 10.3f, 10.3g, and 10.3h were obtained by EBI treatment. Figure 10.3h shows the fibers uniformly isolated, compared with Figure 10.3b, since most of its lignin is removed. Thus, based on the images shown, this proves that the EBI treatment was able separate pure cellulose fibers from a bundle of kenaf fibers via hot water and bleaching processes (Shin et al. 2012). Figure 10.4 showed the SEM images of raw kenaf bast fibers and bleached cellulose fibers.

Kenaf bast fibers show a higher concentration of cellulose versus core fibers in the range of up to 69.2%, which are also higher than other abundant natural fibers such as oil-palm (Khalid et al. 2008). Cellulose is extracted from kenaf fibers using the process of chlorination and mercerization (Tawakkal et al. 2012). The most common

**FIGURE 10.4** Images of SEM of raw kenaf bast fiber and the bleached cellulose fibers: (a) raw kenaf bast fiber; (b) bleached cellulose fiber by alkali treatment; (c), (d), (e), (f), (g), and (h) bleached cellulose fibers obtained by EBI treatment.

method is mercerization or alkali treatment, where the natural fibers are subjected to a strong basic aqueous solution to induce swelling and remove waxy materials, lignin, hemicellulose, and other impurities (Goda et al. 2006).

Tawakkal et al. (2012) prepared α-cellulose and kenaf-derived cellulose (KDC) by two major steps of chlorination and mercerization. The first process of chlorination was used to produce holocellulose, then lignin was eliminated (delignification). During delignification, the color of kenaf bast fibers changed from light brown to white. Lastly, the holocellulose was converted to cellulose by the mercerization process.

## 10.4 CELLULOSE KENAF FIBER-REINFORCED THERMOPLASTIC COMPOSITES

Biocomposites are defined as composite materials prepared from natural fibers and petroleum-derived nondegradable polymers of thermoplastics, such as PP and polyethylene PE (Mohanty et al. 2005). Regarding to the improvement of biocomposites for packaging application, the chance of using kenaf cellulose (KC) was studied in the production of low-density polyethylene (LDPE)/KC/polyethylene glycol (PEG) biocomposites. The result proved that the mechanical properties of the biocomposites slightly decreased as the content of KC increased. Nevertheless, a good homogeneity takes place between samples with added PEG. The addition of PEG enhanced the resistance of thermal of these biocomposites (Tajeddina et al. 2010).

Polylactic acid (PLA) is one of the famous thermoplastics which are used in packaging materials and has similar mechanical properties to other petroleum-based polymers, like PE, polystyrene (PS), PP, and polyethylene terephthalate (PET) (Auras et al. 2004). Moreover, PLA has a some benefits such as biodegradability, the ability to seal at low temperatures, low gas emission, and renewability (Tawakkal et al. 2012). KDC-filled PLA composites are prepared using melt blending and compression molding to upgrade the properties of PLA by presenting the natural cellulose of kenaf fibers. Figure 10.5 illustrates the possible interaction between KDC and PLA. Excellent interfacial adhesion between KDC and PLA occurred due to the presence of hydrogen bonds in composites that resulted from the interaction of hydroxyl groups in kenaf fibers (Bax and Mussing 2008) with carbonyl groups of ester of PLA (Garlotta et al. 2003). The results obtained showed that the mechanical and physical properties of KDC/PLA composites were enhanced than those of the industrial neat PLA polymers (Tawakkal et al. 2012). Tabulation data of the mechanical properties of KDC/PLA composites are shown in Table 10.2.

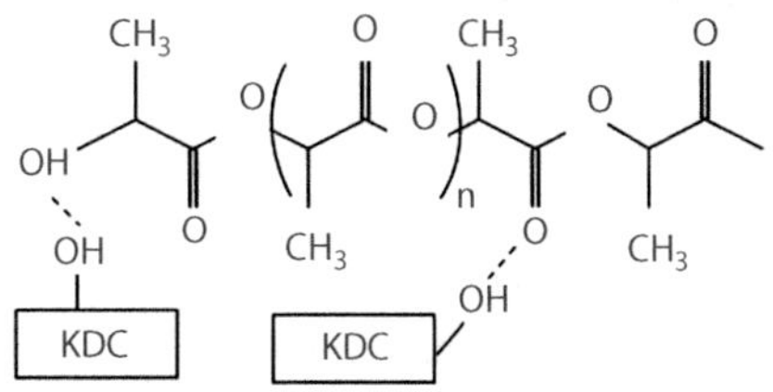

**FIGURE 10.5** Interaction between KDC and PLA. (From Taib, R.M. et al., *Polymer Composites, 31* (7), 1–10, 2009.)

**TABLE 10.2**
Mechanical Properties of KDC/PLA Composites

| Material (%) | Density (g/cm3) | Flexural Strength (MPa) | Flexural Modulus (Gpa) | Impact Strength (J/m) | Elongation at Break (Mpa) |
|---|---|---|---|---|---|
| Neat PLA | 1.26 | 63.4 | 3.9 | 29.9 | 12.3 |
| 10% | 1.27 | 91.8 | 3.9 | 33.4 | 9.0 |
| 20% | 1.30 | 98.8 | 5.5 | 33.9 | 8.7 |
| 30% | 1.32 | 95.3 | 5.4 | 35.3 | 8.4 |
| 40% | 1.35 | 97.2 | 6.6 | 34.8 | 9.0 |
| 50% | 1.39 | 77.3 | 6.4 | 29.5 | 8.7 |
| 60% | 1.41 | 68.2 | 8.4 | 27.4 | 9.1 |

*Source:* Tawakkal, I.S.M.A. et al., *BioResources*, 7(2), 1643–1655, 2012.

Alkaline treatment was used to expose more cellulose on the surface of the fibers to improve the linkage between natural fiber and matrix material (Li et al. 2007). Silane treatment was generally used to improve the bonding between the matrix, since silane breaks down into silanol and alcohol in the presence of water. Alcohol groups in the cell wall of natural fibers react with silanol by forming stable covalent bonds (Agrawal et al. 2000). The benefits of using PP as a matrix are its low cost and comparatively low processing temperature, which is required due to low thermal stability of natural fibers (Maya et al. 201). Asumani et al. (2012) studied the combination of both treatments of alkali-silane (three-aminopropyltriethoxysilane) on kenaf-based cellulose fiber-reinforced polypropylene composites, which significantly enhanced the tensile and flexural properties and was notably better than those gained from either alkali or silane treatment alone.

## 10.5 CELLULOSE KENAF FIBER-REINFORCED THERMOSET COMPOSITES

Thermoset polymer composites have been most extensively investigated due to their massive of application by researchers all around the world (Akil et al. 2011)

Mercerization (alkali treatments) was particularly used to enhance the interfacial linkage strength between ligocellulosic fibers and thermoset resins (Kalia et al. 2009). This method was used to remove hemicellulose and lignin and increase the cellulose content on the surface of the fibers. It has been proven that the alkaline treatment enhances the mechanical properties due to the improvement of the properties of the interfacial adhesive (Fiore et al. 2015). Fiore et al. (2015) pretreated kenaf fibers with sodium hydroxide in different periods of time (48 h and 144 h), and a SEM micrograph of untreated and treated fibers was taken after 48 h and 144 h, shown in Figure 10.6. These cellulose kenaf fiber-reinforced epoxy resin composites showed that they develop higher moduli than neat resin due to the improvement of fiber–matrix compatibility.

Unsaturated resins, such as polyester resins, are used in sheet and bulk molding compounds, and are utilized in a variety of industrial applications such as automotive,

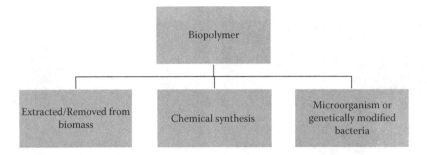

**FIGURE 10.6** Classification of biopolymer based on the origin and production.

aircraft, watercraft, building panels, furnishings, appliances, and fixtures for home construction. Since polyester resins are lightweight and weather resistant, it is widely used as stainless steel in numerous applications. Bing Yang (2014), studied the kenaf based cellulose fiber-reinforced composites with polyester, as well as the effect of glass fiber reveiled that the tensile modulus increased due to reinforcement of kenaf-based cellulose in polyester and the efficiency of stress transfer along the kenaf interface. This resulted in a strong interaction between polyester and kenaf fibers. Table 10.3 presents the modulus predicted with polyester, kenaf fibers, and glass fibers.

Epoxy has been commonly selected as the matrix for hybrid materials due to its unique properties with other thermoset resins, its high mechanical properties, and very low creep and shrinkage, which is suitable for large automotive parts such as bumper beams. Additionally, epoxy is low in viscosity and has good flowing properties that enhance linkage between different reinforcement layers, only requiring low manufacturing pressure (Davoodi et al. 2010). Shalwan, and Yousif (2014) investigated the characteristics of kenaf–epoxy composites exposed to high temperatures using an alkalization process. The SEM analysis of untreated and treated kenaf–epoxy composites are shown in Figure 10.7. Since the cellulose of kenaf fibers were gained from this treatment, the results show the addition of kenaf fibers to the epoxy, slightly enhancing both the charring and thermal stability of the samples.

**TABLE 10.3**
**Modulus of Different Concentration of Polyester, Kenaf Fibers, and Glass Fibers**

|  | Polyester | 40% Kenaf | 40% Kenaf | 60% Kenaf | 50% Glass |
|---|---|---|---|---|---|
| Tensile strength (MPa) | 13.67±0.74 | 15.15±0.62 | 15.3±0.85 | 17.32±0.89 | 15.08±1.22 |
| Strain at break (%) | 0.659±0.098 | 0.333±0.038 | 0.321±0.048 | 0.319±0.049 | 0.146±0.065 |
| Young's modulus (Gpa) | 2.381±0.16 | 4.922±0.37 | 5.153±0.52 | 5.944±0.54 | 13.443±0.89 |
| Density (kg/m$^3$) | 1370 | 1282 | 1260 | 1238 | 1960 |
| Specific modulus ($10^3$m$^2$s$^{-2}$) | 1.74±0.17 | 3.84±0.29 | 4.09±0.41 | 4.81±0.44 | 6.86±0.45 |
| Specific strength ($10^3$m$^2$s$^{-2}$) | 9.98±0.54 | 11.82±0.48 | 12.15±0.67 | 13.99±0.72 | 7.69±0.62 |

*Source:* Yang, B. et al., *Composites Part B: Engineering*, 56, 926–933, 2014.

**FIGURE 10.7** SEM analysis of untreated and treated kenaf–epoxy composites.

## 10.6 CELLULOSE KENAF FIBER-REINFORCED BIOPOLYMERS

Bio-based polymers, also known as biopolymers, are materials derived from renewable resources (Weber et al. 2002). Based on Figure 10.8, bio-based polymers can be divided into three main categories based on their source and production: the polymers directly extracted or removed from biomass such as polysaccharides

# Cellulose-Based Composites from Kenaf Fibers

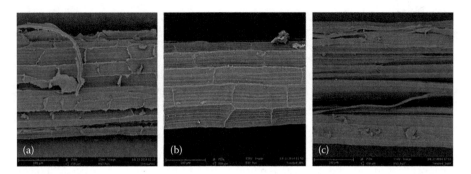

**FIGURE 10.8** SEM micrograph; (a) untreated kenaf fiber, (b) after 48 h in alkaline solution of NaOH, (c) after 144 h.

like starch and cellulose, chitosan–chitin, and proteins such as casein and gluten (Ruban 2009); the polymers produced by conventional chemical synthesis using renewable bio-based polymers, such as is polylactide (Kandemir et al. 2005); and polymers produced by microorganisms or commonly modified bacteria (Cutter 2002).

**FIGURE 10.9** SEM micrograph of untreated and treated kenaf fibers: (a) untreated; (b) acetylated; (c) after block isocyanate treatment; (d) after maleic anhydride treatment; (e) after permanganate treatment. (From Datta, J., Kopczyńska, P., *Industrial Crops and Products*, 74, 566–576. 2015.)

## 10.7 CELLULOSE KENAF FIBER-TREATED COMPOSITES

The hydroxyl group presented in cellulose and lignin were activated by the chemical treatment of fibers. Additionally, chemical treatment introduced new moieties that efficiently connected with polymer matrices (Kalia et al. 2009). Swelling occurred when the alkali reacted with the natural fiber due to the natural crystalline structure of cellulose loosened up and transformed into natural cellulose and cellulose-II. Cellulose-II gave a stabler structure than natural cellulose (John and Anandjiwala 2008) due to the alkali affect and degree of swelling. The presence of the hydroxyl groups made the natural fiber ready for surface modification. Because the chemical treatment, wetting, adhesion, and porosity of the fibers were enhanced, it also enhanced their composite properties (Venkateshwaran et al. 2013). Alavudeen et al. (2015) reported that banana–kenaf fiber-reinforced hybrid polyester is treated using alkali (NaOH) and sodium lauryl sulfate (SLS) in order to give the fibers extra mechanical strength via improved interfacial linkage.

Datta and Kopczyńska (2015) reported that treated kenaf fiber-reinforced polyurethane composites presented enhancement in tensile strength, stiffness, and flexibility, as well as lower water uptake for composites with 10% fiber loading. The surface and linkage of fibers were investigated by using SEM. The SEM resulted in different fiber surfaces after the chemical treatment of acetylation, blocked isocyanate, maleic anhydride, and permanganate treatment. Figure 10.9 shows the SEM

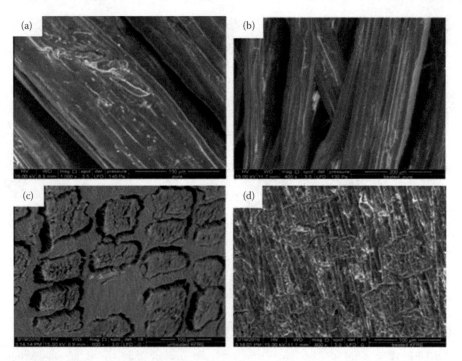

**FIGURE 10.10** SEM analysis of: (a) Untreated kenaf fiber (b) Treated kenaf fiber (6% of NaOH) (c) untreated of KFRE (d) Treated of KFRE. (From Yousif B.F. et al., *Material Design*, 40, 378–385, 2012.)

micrograph of untreated and treated kenaf fibers. In conclusion, the modification of the surface of kenaf fibers changed the mechanical properties.

In another investigation on kenaf-based cellulose done by Yousif et al. (2012), kenaf fibers were treated with 6% of NaOH. The flexural properties of undirectional long kenaf-reinforced epoxy (KFRE) were investigated. The output of this study showed that reinforcement of epoxy with treated kenaf fibers raised the flexural strength of the composite by 36%, while the untreated kenaf fibers initiated 20% of the enhancement, since the high improvement of the alkaline treatments on the interfacial likage of the fibers and the porosity of the composite precluded the debonding, tearing, detachments, and pullout of fibers. SEM was used to analyze the surface morphology of untreated and treated KFRE. Comparatively, the treated KFRE surface became rougher and fibrillated than the untreated KFRE, as shown in Figure 10.10. Additionally, the fibers split into fine cellulose fibers. Thus, the modification of kenaf fibers with 6% of NaOH changed the kenaf fiber bundles into fine cellulose fibers which permitted the epoxy resin to go through and cause high interlocking of the fiber to the matrix.

## REFERENCES

Abdul Khalil, H.P.S., A.F. Ireana Yusra, A.H. Bhat, and M. Jawaid. (2010). Cell wall ultrastructure, anatomy, lignin distribution, and chemical composition of Malaysian cultivated kenaf fibe. *Journal of Industrial Crops and Products* 31, 113–121.

Agrawal, R., N.S. Saxena, K.B. Sharma, S. Thomas, & M. Sreekala. (2000). Activation energy and crystallization kinetics of untreated and treated oil palm fibre reinforced phenol formaldehyde composites. *Materials Science and Engineering: A* 277 (1-2), 77–82.

Aji I.S., S.M., Sapuan, E.S. Zainudin, and K. Abdan. (2009). Kenaf fibres as reinforcement for polymeric composites: A review. *International Journal Mechanical Material Engineering* 4, 239–248.

Akil, H.M., M.F. Omar, A.A.M. Mazuki, S. Safiee, Z.A.M. Ishak, and A. Abu Bakar School. (2011). Kenaf fiber reinforced composites: A review. *Materials and Design*, 32 (8–9), 4107–4121.

Alavudeen, A., N. Rajini, S. Karthikeyan, M. Thiruchitrambalam, and N. Venkateshwaren. (2015). Mechanical properties of banana/kenaf fiber-reinforced hybrid polyester composites: Effect of woven fabric and random orientation. *Materials and Design* 66, 246–257.

Anuar, H. and A. Zuraida. (2011). Improvement in mechanical properties of reinforced thermoplastic elastomer composite with kenaf bast fibre. *Journal of Composite Part B: Engineering* 42 (3), 462–465.

Asumani, O.M.L., R.G. Reid, & R. Paskaramoorthy. (2012). The effects of alkali–silane treatment on the tensile and flexural properties of short fibre non-woven kenaf reinforced polypropylene composites. *Composites Part A: Applied Science and Manufacturing* 43 (9), 1431–1440.

Auras, R., B. Harte, and S. Selke. (2004). An overview of polylactides as packaging materials. *Macromolecular Bioscience* 4 (9), 835–864.

Baillie, C. (2005). *Green Composites: Polymer Composites and the Environment.* 308: CRC Press.

Bax, B. and J. Mussing. (2008). Impact and tensile properties of PLA/cordenka and PLA/flax composites. *Composites Science and Technology* 68, 1601–1607.

Bella, G. Di., V. Fiore, G. Galtieri, C. Borsellino, and A. Valenza. (2014). Effects of natural fibres reinforcement in lime plasters (kenaf and sisal vs. Polypropylene). *Journal of Construction and Building Materials* 58, 159–165.

Cutter, C.N. (2002). Microbial control by packaging: A review. *Critical Reviews in Food Science and Nutrition* 42 (2), 151–161.

Datta, J., & P. Kopczyńska. (2015). Effect of kenaf fibre modification on morphology and mechanical properties of thermoplastic polyurethane materials. *Industrial Crops and Products* 74, 566–576.

Davoodi, M.M., S.M., Sapuan, D. Ahmad, A. Ali, A. Khalina, and M. Jonoobi. (2010). Mechanical properties of hybrid kenaf/glass reinforced epoxy composite for passenger car bumper beam. *Journal of Material Design* 31, 4927–4932.

Elsaid, A., M. Dawood, R. Seracino, and C. Bobko. (2011). Mechanical properties of kenaf fiber reinforced concrete. *Construction Building Material* 25 (4), 1991–2001.

El-Shekeil, Y.A., S.M. Sapuan, K. Abdan, and E.S. Zainudin. (2012). Influence of fiber content on the mechanical and thermal properties of Kenaf fiber reinforced thermoplastic polyurethane composites. *Materials and Design* 40, 299–303.

Fiore, V., G. Di Bella, and A. Valenza. (2015). The effect of alkaline treatment on mechanical properties of kenaf fibers and their epoxy composites. *Composite: Part B*, 68, 14–21.

Garlotta, D., W. Doane, R. Shogren, J. Lawton and J.L. Willett. (2003). Mechanical and thermal properties of starch-filled poly(D,L-lactic acid)/poly(hydroxy ester ether) biodegradable blends. *Journal of Applied Polymer Science* 88, 1775–1786.

Goda, K., M.S. Sreekala, A. Gomes, T. Kaji, and J. Ohgi. (2006). Improvement of plant based natural fibers for toughening green composites-Effect of load application during mercerization of ramie fibers. *Composites Part A: Applied Science and Manufacturing* 37 (12), 2213–2220.

Holbery, James, and Dan Houston. (2006). Natural fiber reinforced polymer composites in automotive applications. *Low-Cost Composites in Vehicle Manufacture* 58 (11), 80–86.

John, M.J., & R.D. Anandjiwala. (2008). Recent developments in chemical modification and characterization of natural fiber-reinforced composites. *Polymer composites* 29 (2), 187–207.

Jonoobi, M., J. Harun, A. Shakeri, M. Misra, and K. Oksman. (2009). Chemical composition, crystallinity, and thermal degradation of bleached and unbleached kenaf bast (Hibiscus cannabinus) pulp and nanofibers. *Journal of BioResources* 2 (4), 626–639.

Kalia, S., B.S. Kaith, and I. Kaur. (2009). Pretreatments of natural fibers and their application as reinforcing material in polymer composites: A review. *Polymer Engineering Science* 49, 1253–1272.

Kandemir, N., A. Yemenicioğlu, Ç. Mecitoğlu, Z.S. Elmacı, A. Arslanoğlu, M.Y. Göksungur, and T. Baysal. (2005). Production of antimicrobial films by incorporation of partially purified lysozyme into biodegradable films of crude exopolysaccharides obtained from *Aureobasidium pullulans fermentation*. *Food Technology and Biotechnology* 43, 343–350.

Karimi, Samaneh., Paridah Md. Tahir, Ali Karimi, Alain Dufresne, and Ali Abdulkhani. (2014). Kenaf bast cellulosic fibers hierarchy: A comprehensive approach from micro to nano. *Journal of Carbohydrate Polymers* 101, 878–885.

Keshk, S., W. Suwinarti, and K. Sameshima. (2006). Physicochemical characterization of different treatment sequences on kenaf bast fiber. *Journal of Carbohydrate Polymer* 65 (2), 202–206.

Khalid, M., C.T. Ratnam, T.G. Chuah, S. Ali, and T.S.Y. Chong. (2008). Comparative study of polypropylene composites reinforced with oil palm empty fruit bunch and oil palm derived cellulose. *Journal of Materials and Design* 29 (1), 173–178.

Lam, Thi Bach Tuyet, Hori, Keko, and Iiyama, Kenji. (2003). Structural characteristics of cell walls of kenaf (*Hibiscus cannabinus* L.) and fixation of carbon dioxide. *Journal of Wood Science* 49 (3), 255–261.

Lips, Steef J.J., and Jan E.G. van Dam. (2013). Kenaf: A Multi-Purpose Crop for Several Industrial Applications. *Industrial Applications, Green Energy and Technology* 105–143.

Li, X., L.G. Tabil, & S. Panigrahi. (2007). Chemical treatments of natural fiber for use in natural fiber-reinforced composites: A review. *Journal of Polymers and the Environment* 15 (1), 25–33.

Mahjoub, Reza, Jamaludin Mohamad, Abdul Rahman, Mohd Sam, and Sayed Hamid. (2014). Tensile properties of kenaf fiber due to various conditions of chemical fiber surface modifications. *Journal of Construction and Building Materials* 55, 103–113.

Mahjouba, Reza, Jamaludin Mohamad Yatim, Abdul Rahman Mohd Sam, and Mehdi Raftari. (2014). Characteristics of continuous unidirectional kenaf fiber reinforced epoxy composites. *Journal of Material and Design* 24, 640–649.

Maya, Jacob, and Ananjiwala Rajesh. (2008). Recent developments in chemical modification and characterization of natural fiber-reinforced composites. *Polymer Composite* 29, 187–207.

Mazuki, A.A.M., H.M. Akil, S. Safiee, Z.A.M. Ishak, and A.A. Bakr. (2011). Degradation of dynamic mechanical properties of pultruded kenaf fiber reinforced composites after immersion in various solutions. *Journal of Composites Part B* 42 (1), 71–76.

Mohanty, Amar K., Manjusri Misra, and Lawrence T. Drzal. (2005). *Natural Fibers, Biopolymers, and Biocomposites*. Broken Sound Parkway: CRC Press.

Mohd Edeerozey, A.M., H. Md Akil, A.B. Azhar, and M.I. Zainal Ariffin. (2007). Chemical modifications of kenaf fibers. *Journal of Material Letters* 61, 2023–2025.

Mossello, Ahmad Azizi, Jalaluddin Harun, Paridah Md Tahir, Hossein Resalati, Rushdan Ibrahim, Seyed Rashid Fallah Shamsi, and Ainun Zuriyati Mohmamed. (2010). A Review of Literatures Related of Using Kenaf for Pulp Production (Beating, Fractionation, and Recycled Fiber). *Modern Applied Science* 4 (9), 21–29.

Rassiah, K. and M.M.H.M. Ahmad. (2013). A review on mechanical properties of bamboo fiber reinforced polymer composite. *Australian Journal of Basic and Applied Sciences* 7, 247–253.

Ruban, S.W. (2009). Biobased Packaging–Application in meat industry. *Veterinary World*, 2(2), 79–82.

Saba, N., M.T. Paridah, and M. Jawaid. (2015). Mechanical properties of kenaf fibre reinforced polymer composite: A review. *Construction Building Material* 76, 87–96.

Salleh, Fauzani Md., Aziz Hassan, Rosiyah Yahya, and Ahmad Danial Azzahari. (2014). Composites: Part B Effects of extrusion temperature on the rheological, dynamic mechanical and tensile properties of kenaf fiber/HDPE composites. *Journal of Composites: Part B*, 58, 259–266.

Shalwan, A., & B.F. Yousif. (2014). Investigation on interfacial adhesion of date palm/epoxy using fragmentation technique. *Materials and Design* 53, 928–937.

Sheldon, A. (2014). *Preliminary Evaluation of Kenaf as a Structural Material. A Thesis Presented to the Graduate School of Clemson University In Partial Fulfillment of the Requirements for the Degree Master of Science Civil Engineering*, 1–57.

Shin, Hye Kyoung., Joon Pyo Jeun, Hyun Bin Kim, and Phil Hyun Kang. (2012). Isolation of cellulose fibers from kenaf using electron beam. *Radiation Physics and Chemistry* 81 (8), 936–940.

Siro, I. and D. Plackett. (2010). Microfibrillated cellulose and new nanocomposite materials, a review. *Journal of Cellulose* 17, 459–494.

Siti Yasmine Zanariah Zainuddin, Ishak Ahmad, Hanieh Kargarzadeh, Ibrahim Abdullah, and Alain Dufresne. (2013). Potential of using multiscale kenaf fibers as reinforcing filler in cassava starch-kenaf biocomposites. *Journal of Carbohydrate Polymers* 92 (2), 2299–2305.

Sixta, H. (2006). *Handbook of Pulp*. Germany: WILEY-VCH Verlag GmbH &Co. KGaA.

Summerscales, J., N.P.J. Dissanayake, A.S. Virk, and W. Hall. (2010). A review of bast fibres and their composites. Part 1. Fibres as reinforcements. *Composites Part A: Applied Science and Manufacturing* 41 (10), 1329–1335.

Tahir, Paridah Md., Amel B. Ahmed, Syeed O.A. SaifulAzry, and Zakiah Ahmed. (2011). Retting Process of Some Bast Plant Fibres and Its Effect on Fibre Quality: A Review. *BioResources* 6 (4), 5260–5281.

Taib, R.M., S. Ramarad, Z.A.M. Ishak, and M. Todo. (2009). Properties of kenaf fiber/polylactic acid biocomposites plasticized with polyethylene glycol. *Polymer Composites* 31 (7), 1–10.

Tajeddina, Behjat, Russly Abdul Rahman, and Luqman Chuah Abdulah. (2010). The effect of polyethylene glycol on the characteristics of kenaf cellulose/low-density polyethylene biocomposites. *International Journal of Biological Macromolecules* 47, 292–297.

Tajeddin, B., R.A. Rahman, and L.C. Abdullah. (2009). Mechanical and morphology properties of kenaf cellulose/LDPE biocomposites. *American-Eurasian Journal of Agriculture and Environment Science* 5 (6), 777–785.

Tawakkal, Intan Syafinaz M.A., Rosnita A. Talib, Khalina Abdan, and Chin Nyuk Ling. (2012). Mechanical and physical properties of kenaf- derived cellulose (kdc)-filled polylactic acid (pla) composites. *BioResources* 7 (2), 1643–1655.

Venkateshwaran, N., A. Elaya Perumal, and D. Arunsundaranayagam. (2013). Fiber surface treatment and its effect on mechanical and visco-elastic behaviour of banana/epoxy composite. *Materials and Design* 47, 151–159.

Weber, C.J., V. Haugaard, R. Festersen, and G. Bertelsen. (2002). Production and applications of biobased packaging materials for the food industry. *Journal of Food Addatives Contaminants* 19 (4), 172–177.

Yang, B., M. Nar, D.K. Visi, M. Allen, B. Ayre, C.L. Webber III, H. Lu, and N.A. D'Souza. (2014). Effects of chemical versus enzymatic processing of kenaf fibers on poly (hydroxybutyrate-co-valerate)/poly (butylene adipate-co-terephthalate) composite properties. *Composites Part B: Engineering* 56, 926–933.

Yousif, B.F., A. Shalwan, C.W. Chin, and K.C. Ming. (2012). Flexural properties of treated and untreated kenaf/epoxy composites. *Material Design* 40, 378–385.

# 11 Development and Characterization of Kenaf Nanocomposites

*J. Sahari, M.A. Maleque, and M.L. Sanyang*

## CONTENTS

11.1 Introduction: Kenaf Fibers ........................................................................... 185
11.2 Extraction of Nanocellulose from Kenaf Fibers ........................................... 188
    11.2.1 Extraction of Cellulose Nanofiber ..................................................... 188
    11.2.2 Other Extraction Methods of Nanofibrils .......................................... 192
11.3 Nanofiber Reinforced with Composites ....................................................... 194
References .......................................................................................................... 199

## 11.1 INTRODUCTION: KENAF FIBERS

Kenaf is a non-wood lignocellulose fiber crop and belongs to the genus *Hibiscus*. It is also known as a cellulosic source with economic and ecological advantages. Kenaf shows low density, nonabrasiveness during processing, high specific mechanical properties, and biodegradability. Historically, kenaf stems were used as a source of fiber for ropes and sacks, but nowadays, kenaf fibers have been known as reinforcing agents for polymer matrix composites (Liu et al. 2007). There are a few new applications for kenaf, including absorbents, building materials, paper products, and animal feeds. Besides that, using kenaf as a raw material is a selective choice for wood in pulp and paper industries to avoid deforestation. National kenaf research and development programs have been formed in Malaysia after realizing the various possibilities of commercially exploitable products derived from kenaf. The main objective of these programs was to develop kenaf as a possible new industrial crop for Malaysia. Table 11.1 shows the mechanical properties of kenaf fibers and Figure 11.1 shows kenaf plantation and kenaf fibers. A transmission electron micrograph of ultrathin sections of kenaf fibers is shown in Figure 11.2.

Sustainable properties of kenaf include (Zampaloni et al. 2007):

    a. Absorption of carbon dioxide from the air at a high rate
    b. Absorption of nitrogen and phosphorous from the soil

## TABLE 11.1
### Mechanical Properties of Kenaf Fibers

| Source | Density (g/cm³) | Tensile Strength (MPa) | Tensile Modulus (GPa) | Elongation (%) |
|---|---|---|---|---|
| Cicala et al. 2009 | – | 692 | 10.94 | 4.3 |
| Akil et al. 2011 | – | 930 | 53 | 1.6 |
| Saheb and Jog 1999 | 1.45 | 930 | 53 | 1.6 |
| Parikh et al. 2002 | 1.4 | 284–800 | 21–60 | 1.6 |
| Cheung et al. 2009 | – | 295–1191 | 2.86 | 3.5 |
| Rassmann et al. 2010 | 1.5 | 350–600 | 40 | 2.5–3.5 |
| Ribot et al. 2011 | 0.75 | 400–550 | – | – |
| Yousif et al. 2012 | 0.6 | – | – | – |
| Graupner et al. 2009 | 0.749 | 223–624 | 11–14.5 | 2.7–5.7 |
| Jawaid and Abdul Khalil 2011 | 1.2 | 295 | – | 3–10 |

(a)          (b)

**FIGURE 11.1** (a) Kenaf Plantation; (b) kenaf fibers. (From Saba, N. et al., *Construction and Building Materials*, 76, 87–96, 2015.)

    c. Recyclability
    d. A light weight which helps enhance the fuel consumption and emissions, especially in the automotive industry

Due to environmental awareness and the international demand for green technology, bionanocomposites are one of many types of composites highly recommended to replace present petrochemical-based materials. The use of natural fibers

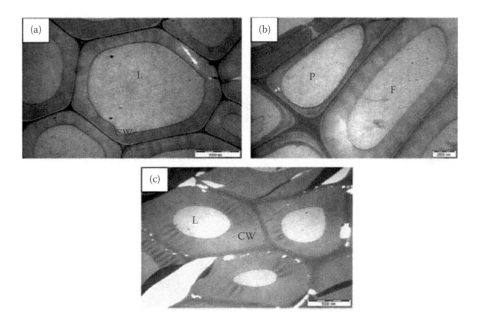

**FIGURE 11.2** Transmission electron micrograph of an ultra-thin section of kenaf fibers: (a) kenaf core fibers showing cell wall and lumen; (b) transverse section of core fibers; (c) transverse section of kenaf bast fibers. Cell wall structure stained with uranyl acetate and lead citrate. Scale bars: 5000 nm (a and c); 2000 nm in (b) cell wall; L, lumen; F, fiber; P, parenchyma. (From Khalil, H.A. et al., *European Polymer Journal*, 37(5): 1037–1045, 2001.)

instead of traditional reinforcement materials, such glass fibers, carbon and talc, and cellulose nanofibers (CNFs) have attracted interest in several applications such as biomedicine, bioimaging, nanocomposites, gas barrier films, and optically transparent functional materials (Siqueira et al. 2010). Instead of extracting CNFs from algae, tunicates, and bacterial cellulose, CNFs also can be extracted from natural plant cell walls (Klemm et al. 2005). Figure 11.3 shows the cellulose cell wall and microfibril organization (Siqueira et al. 2010).

In the last decade, CNFs showed great potential as nano-reinforcing materials in different polymers influenced by their appealing intrinsic properties including nanoscale dimensions, high surface area, low density and high mechanical strength, renewability, and biodegradability (Abdul Khalil et al. 2012). The properties of fiber-reinforced composites are strongly due to the interactions between the components. The main problems of researchers in this field are mostly related to the highly polar surface of the cellulose fibers, which causes a very low interfacial compatibility with non-polar polymer matrices, moisture uptake, and inter-fiber aggregation by hydrogen bonding (Cunha and Gandini 2010). Several techniques have been studied to overcome these problems such as esterification (Mastsumura et al. 2000), silylation (Gousse et al. 2004), and acetylation (Hu et al. 2011). These modifications are mostly involved specific surface treatments of the fibers for decreasing the CNF's hydrophilicity (Hadi et al. 2015).

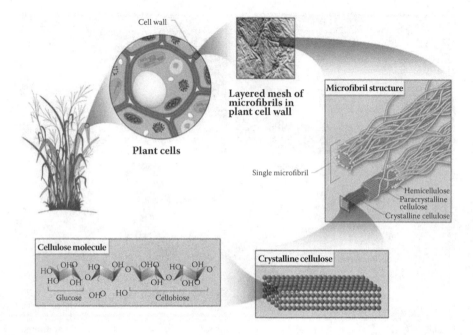

**FIGURE 11.3** Scheme of cellulose cell wall and microfibril organization. (From Siqueira, G. et al., *Polymers* 2: 728–765, 2010.)

## 11.2 EXTRACTION OF NANOCELLULOSE FROM KENAF FIBERS

### 11.2.1 Extraction of Cellulose Nanofiber

Usually, CNFs are extracted from lignocellusic plants, which have a high modulus and high surface area. These CNFs are more suitable as reinforcers for polymer matrices (Petersson et al. 2007). However, for developing composites, they are still restricted due to hydrophilicity of the CNFs as well as the formation of irreversible aggregates when dried (Siró and Plackett 2010). Moreover, the uniform dispersion of the CNFs in polymer is strongly influenced by their high surface energy and the presence of hydroxyl groups on the CNF's surface. To solve these problems, various chemical modifications have been investigated.

Among the many chemical modifications, acetylation is one of the most promising processes because it can improve the dispersibility of nanofibers in polymer matrix (Ashori et al. 2014) and the dimensional stability of the final composition (Khalil et al. 2001). During the acetylation, the chemical will react with the OH groups on the cellulose, thereby modifying the hydrophilic surface of the cellulose to become hydrophobic. The acetylation of cellulose depends on the accessibility and susceptibility of the OH-groups in the amorphous and crystalline regions within the cellulose polymer chain (Sassi and Chanzy 1995). Then, to extract nanofibers, mechanical isolation processes are used, as briefly shown in the Figure 11.4. Basically, mechanical processes consist of disintegration, refining, cryogenic grinding, and high-pressure homogenization.

# Development and Characterization of Kenaf Nanocomposites

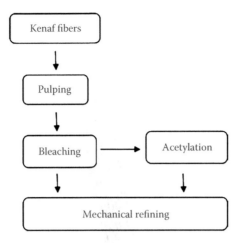

**FIGURE 11.4** Schematic of the isolation process of nanofibers and acetylated nanofibers. (From Jonoobi, M. et al., *Cellulose* 17(2): 299–307, 2010a.)

According Jonoobi et al. (2010a), CNFs were isolated from kenaf fibers using the acetylated method, which acetic anhydride (1:20 by weight) and 5 wt% pyridine as catalyst being used under reflux at 100 °C for 4 h. Finally, in order to remove unreacted acetic anhydride and acetic acid by-product, the treated CNFs were extracted using distilled water (2:1 by volume) for 6 h. Then, to obtain aqueous dispersions of the ACNFs, four cycles of solvent exchange were performed by centrifugation (Babaee et al. 2015). Figures 11.5 and 11.6 show the chemical reaction that occured during the acetylation process and hypothesis of the nanofibers' acetylation (Jonoobi et al. 2010a).

Other researchers had extracted cellulose from kenaf bast fibers using soda–anthraquinone (AQ) pulping and a subsequent alkaline-peroxide bleaching procedure. This pulping can decrease the rate of degradation and cause less damage to cellulose chain, a fact that has been attributed to stabilization effect of AQ on the cellulose macromolecules. In addition, AQ was suitable way to block the active groups of the cellulose and reduce the rate of oxidation. First, kenaf fibers were cut into 2–3 cm lengths. After that, 300g of oven-dried, short kenaf fibers were cooked in a digester (MK model, USA) with 15% of NaOH and 0.1% of AQ.

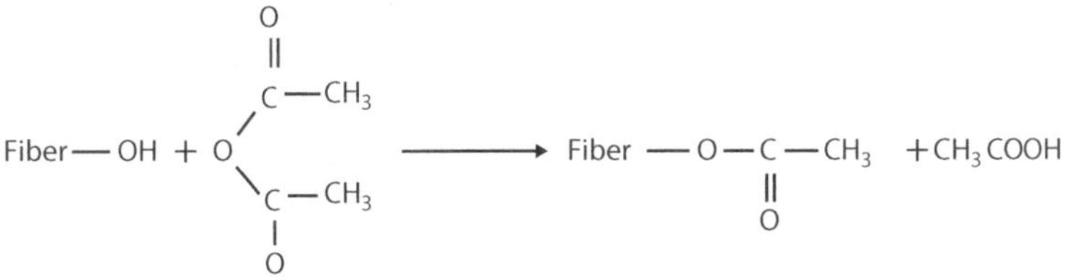

$$\text{Fiber—OH} + \underset{\underset{O}{|}}{\overset{O}{\underset{C-CH_3}{\overset{\|}{C-CH_3}}}}\!\!\!\!\!\!O \longrightarrow \text{Fiber}-O-\underset{O}{\overset{\|}{C}}-CH_3 + CH_3COOH$$

**FIGURE 11.5** Chemical reaction of acetic anhydride with cellulose. (From Jonoobi, M. et al., *Cellulose* 17(2): 299–307, 2010a.)

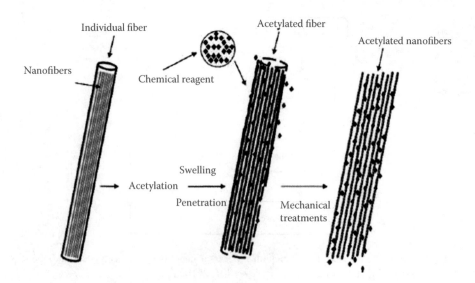

**FIGURE 11.6** Hypothesis of the nanofibers' acetylation. (From Jonoobi, M. et al., *Cellulose* 17(2): 299–307, 2010a.)

The cooking liquor to kenaf fiber ratio was 7:1 with 1% AQ at 160 °C for 2 h. The maximum cooking temperature was set to 160 °C, and this temperature was reached after 60 min. Another 45 min of cooking was carried out at this temperature. Table 11.2 shows the conditions of the NaOH–AQ pulping process (Jonoobi et al. 2009).

Besides Jonoobi, Yalda et al. (2015) also used soda–AQ pulping for preparation of kenaf bast CNF and then proceed to use an alkaline–peroxide bleaching process. First they cut kenaf fibers into 2–3 lengths. Then pulping was conducted with the following conditions: 200g of fiber, 19.4% NaOH, the cooking liquor to kenaf fiber ratio was 7:1, and 1.0% AQ at 160 °C for 2 h. Subsequently, they rinsed the pulp with water to remove the chemicals used in the digester. The yield of pulping

**TABLE 11.2**
**Conditions of the NaOH–AQ Pulping Process**

| Cooking Process | NaOH–AQ |
|---|---|
| NaOH | 15% |
| AQ (anthraquinone) | 0.1% |
| Liquor-to-fiber ratio | 7:1 |
| Maximum temperature | 160 °C |
| Time to maximum temperature | 60 min |
| Time at maximum temperature | 45 min |

# Development and Characterization of Kenaf Nanocomposites

process was about 56%. Then, a bleaching treatment was performed to eliminate any residual lignin using 3% H2O2, 3% NaOH, and 0.5% MgSO4 at 80 °C for 2 h (Ashori et al. 2006). Finally, 10 g of the bleached kenaf bast fibers were hydrolyzed using 1.5 M HCl at 80 °C for 2 h. The hydrolyzed fibers were then rinsed in distilled water to neutralize the pH, filtered using a membrane filter (Nylafo membrane disc filter 0.2_m, Pall Corporation, Malaysia), and after that, oven-dried at 60 °C for 24 h. A HCl hydrolysis was implemented in this study as a pretreatment prior to the homogenization process in order to reduce the size of fibers to prevent the clogging the homogenization's nozzle, facilitating disintegration, removing amorphous sections from fibers, and for disintegrating CNF from the kenaf bast cell wall (Pan et al. 2013).

In processing the nanocellulose from kenaf core, a chemical-mechanical treatment was used as reported by Surip et al. (2012). Kenaf fibers were preliminarily soaked with sodium hydroxide solution with a 6% w/w ratio for hours. After that, fibers were washed with distilled water. Then, they were treated by hydrochloric acid (HCl). Mechanical treatment was conduct by separating the fiber bundles into single fibers. For this purposed, treated fibers were crushed using cryogenic crusher with the presence of nitrogen for further reduction into nanoscale fibers. Finally, the fibers were dried in the oven. Figures 11.7 and 11.8 show the scanning electron microscope (SEM) of separated and treated kenaf nanocellulose fibers with different magnifications. An illustration of the hydrogen bonding of nanocellulose is shown in Figure 11.9.

Jonoobi et al. (2010c) isolated cellulose nanofibers from kenaf core fibers by employing chemo-mechanical treatments. Nanofiber morphologies were examined in every stage. The transmission electron microscopy results showed that most of the nanofibers isolated from the kenaf core fibers had diameters in the range of 20–25 nm. Chemical analysis of the fibers after each stage of the treatment indicated an increase in their cellulose contents and a decrease in their lignin and hemicellulose contents relative to the corresponding contents before each respective treatment.

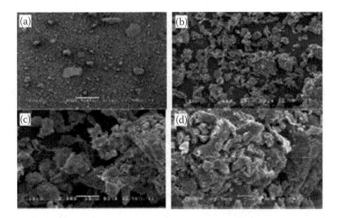

**FIGURE 11.7** SEM micrograph of separated kenaf nanocellulose under different magnification: (a) 40×; (b) 600×; (c) 2000×; and (d) 3000×. (From Surip, S.N. et al., *Advanced Materials Research* 488–489: 72–75, 2012.)

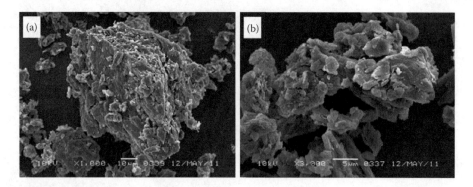

**FIGURE 11.8** SEM micrograph of treated kenaf nanocellulose under (a) 1000× magnification and (b) 3000× magnification. (From Surip, S.N. et al., *Advanced Materials Research* 488–489: 72–75, 2012.)

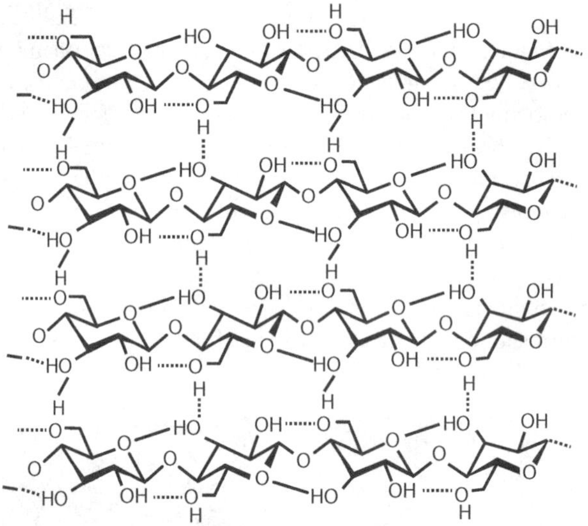

**FIGURE 11.9** Illustration of hydrogen bonding of nanocellulose. (From Surip, S.N. et al., *Advanced Materials Research* 488–489: 72–75, 2012.)

### 11.2.2 Other Extraction Methods of Nanofibrils

Saito et al. (2006) accomplished an oxidation pretreatment of cellulose by applying 2,2,6,6 tetramethylpiperidine-1-oxyl (TEMPO) radicals before mechanical treatment. The TEMPO–mediated oxidation was used to oxidize softwood and hardwood celluloses. The TEMPO-oxidized cellulose fibers were transformed by transparent dispersions in water, which consisted of individual nanofibers 3–4 nm in width. Films were then prepared from the TEMPO–oxidized cellulose nanofibers (TOCN) and characterized from various aspects. AFM images showed that the TOCN film surface consisted of randomly assembled cellulose nanofibers. A systematic diagram of individualization of nanosized plant cellulose fibrils by surface carboxylation using TEMPO catalyst is shown in Figure 11.10. Figure 11.11 shows water absorption of the TPS and its nanocomposites (Babaee et al. 2015).

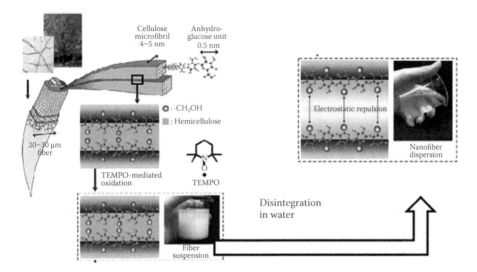

**FIGURE 11.10** Systematic diagram of individualization of nanosized plant cellulose fibrils by surface carboxylation using a TEMPO catalyst. (From Saito, T. et al., *Biomacromolecules* 8(8): 2485–2491, 2007.)

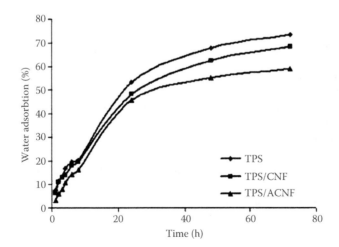

**FIGURE 11.11** Water absorption of the TPS and its nanocomposites. (From Babaee, M. et al., *Carbohydrate Polymers* 132: 1–8, 2015.)

## TABLE 11.3
### Other Extraction Method of Nanofibrils

| Extraction Processes and Methods | References |
|---|---|
| Mechanical treatments, e.g., cryo-crushing grinding | Chakraborty et al. 2005, 2006; Abe et al. 2007, 2009; Abe and Yano 2009, 2010; Nogi et al. 2009 |
| High pressure homogenizing; chemical treatments e.g., acid hydrolysis; biological treatments e.g., enzyme-assisted hydrolysis | Nakagaito and Yano 2004, 2005, 2008a, 2008b; Araki et al. 2000; Elazzouzi-Hafraoui et al. 2007; Liu et al. 2007; Hayashi et al. 2005; Henriksson et al. 2007; Paakko et al. 2007 |
| TEMPO-mediated oxidation on the surface of microfibrils and a subsequent mild mechanical treatment; synthetic and electrospinning methods | Iwamoto et al. 2010; Saito et al. 2009, 2007, 2006; Frenot et al. 2007; Kim et al. 2006; Ma et al. 2005 |
| Ultrasonic technique | Cheng et al. 2010, 2009, 2007; Wang and Cheng 2009, Zhao et al. 2007 |

Cryogenic crushing is an alternative method for producing nanofibers in which fibers are frozen using liquid nitrogen and high shear forces are then applied. By combining the mechanical treatment with certain pretreatments, it is possible to decrease energy consumption. Before mechanical processing, a number of researchers have applied alkaline treatment of fibers in order to disrupt the lignin structure and help to separate the structural linkages between lignin and carbohydrates (Wang and Sain, 2007). Purification by mild alkali treatment results in the solubilisation of lignin, pectins, and hemicelluloses. Oxidation and ezymatic pretreatment are another method. Enzymatic pretreatment reduces energy consumption. Other extraction methods of nanofiber are shown in Table 11.3.

## 11.3 NANOFIBER REINFORCED WITH COMPOSITES

Babaee et al. (2015) reported the use of cellulose nanofibers as the reinforcement for thermoplastic starch (TPS) nanocomposites. The purpose of this study was to investigate the effect of chemical modification of cellulose nanofibers on the biodegradability and mechanical properties of TPS. From Table 11.4, the study revealed that the mechanical properties of composites with the addition of CNFs and ACNFs had significantly increased in tensile strength and showed reduction in the Young's Modulus (MPa) result. The increase in tensile strength and Young's modulus may be due to the formation of a hydrogen-bonded nanofiber network and also nanofiber entanglement (Angles and Dufresne, 2001). Moreover, it also may be caused by the formation of hydrogen bonding resulting in strong interface between two phases (matrix and nanofibers) (Sreekala et al. 2008). In addition, the low tensile modulus and strength of the TPS treated with the ACNFs compared to TPS treated with the CNFs, are the reason for the reduction of nanofibers-to-nanofiber and

# Development and Characterization of Kenaf Nanocomposites

## TABLE 11.4
### Mechanical Properties of the TPS and Its Nanocomposites

| Tensile Strength (MPa) | Young's Modulus (MPa) | Elongation at Break (%) | Materials |
|---|---|---|---|
| 8.6 ± 1 | 16.6 ± 2 | 52 ± 2 | TPS |
| 38.0 ± 3 | 141.0 ± 35 | 27 ± 4 | CNFs/TPS |
| 14.7 ± 2 | 29.6 ± 6 | 50 ± 2 | ACNFs/TPS |

*Source:* Babaee, M. et al., *Carbohydrate Polymers* 132: 1–8, 2015.

nanofiber-to-matrix interactions due to the surface hydrophobicity and also the reduced crystallinity of the nanofiber after acetylation (Jonoobi et al. 2012). Figure 11.12 shows the water absorption of TPS nanocomposites reinforced with CNFs and ACNFs. It shows a significant reduction in water absorption of the TPS with the addition of CNFs and ACNFs. The reduction in water may due to the amorphous and branch structure of the starch with higher accessibility to water than CNFs, while for ACNFs it might be due to the acetylation process.

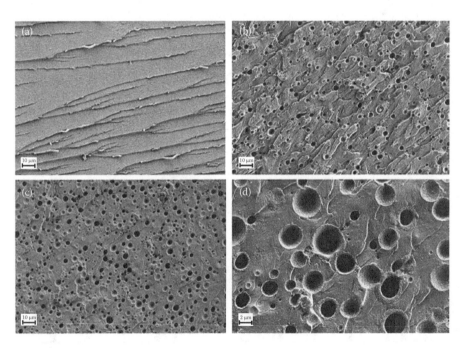

**FIGURE 11.12** FESEM micrographs of neat UPR and LENR-UPR blend: (a) UPR; (b) 1.5 wt%; (c) 4.5 wt%; (d) 4.5 wt% blend at higher magnification. White arrows in (c) show the rubber particles which remains in the matrix after toluene extraction. White arrows in (d) show the crack pinning effect induced by LENR. (From Hanieh, K. et al., *Industrial Crops and Product* 72: 125–132, 2015.)

Hanieh et al. (2015) reported using liquid epoxidized natural rubber (LENR) to modify the unsaturated polyester resin (UPR) and introduce nanocrystals into it to produce toughened polyester cellulose nanocomposites. Cellulose nanocrystals were prepared from the sulfuric hydrolysis of cellulose extracted from kenaf bast fibers. The main purpose of this experiment was to determine the effect of the LENR and CNC content on the mechanical properties and the morphology of the nanocomposites. From Figure 11.12, it was observed that the size of the rubber particles is independent of the rubber content, influenced by the chemical interaction of rubber and matrix. However, in their findings, the nanocomposite blend revealed a different result for tensile strength as observed in Figure 11.13. The highest value of tensile strength was due to the improved compatibility of the LENR–UPR blend matrix via chemical interaction; while at a lower result of tensile modulus, it might be attributed to the LENR's soft segment structure as well as to the intrinsically low modulus of liquid rubbers compared to neat UPR (Seng et al. 2011).

Previous researchers have produced biodegradable composites from cassava starch and kenaf fibers as reinforcing fillers. In this study, kenaf fibers underwent alkalization (treated with NaOH), bleaching (bleached with sodium chlorite and acetic buffer solution), and hydrolysis treatments. Cellulose nanocrystals (CNCs) from kenaf were prepared by acid hydrolysis of the cellulose obtained from kenaf. The objective of this experiment was to observed effect of using multiscale kenaf fibers on biodegradable composites. The work revealed the morphological observation and decrease in size of extracted cellulose and CNCs. It also revealed that the tensile strength and moduli of the biocomposites increased after each treatment was conducted. The highest tensile observed at 8.2 MPa (6% TPCS–CNCs) among the composites for all of the fiber loading. Figure 11.14 shows the TEM

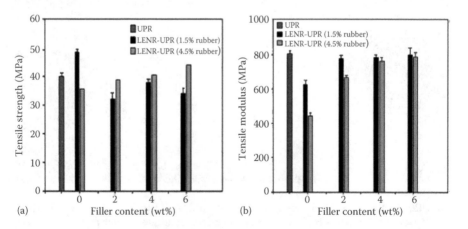

**FIGURE 11.13** Tensile strength and tensile modulus of nanocomposites. (From Hanieh, K. et al., *Industrial Crops and Product* 72: 125–132, 2015.)

# Development and Characterization of Kenaf Nanocomposites

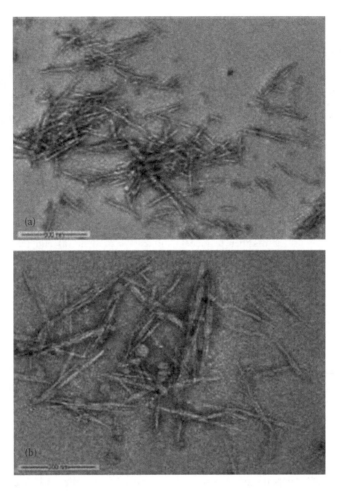

**FIGURE 11.14** TEM micrograph of hydrolyzed kenaf under different magnifications: 13,000× (a) and 35,000× (b). (From Siti Yasmine, Z.Z. et al., *Carbohydrate Polymers* 92: 2299–2305, 2013.)

micrograph of hydrolyzed kenaf under different magnifications and Figure 11.15 shows the effect of the kenaf on the mechanical properties of the biocomposites (Siti Yasmine et al. 2013).

Previous study on cellulose nanofiber (CNF)–reinforced polylatic acid (PLA) prepared by twin screw extrusion were conducted by Jonoobi et al. (2010b). From these studies, the nanofibers were separated from kenaf pulp and were used as reinforcing filler. Two-step processes were conducted to prepare the nanocomposites, which were prepared as a master batch using a solvent mixture then followed by an extrusion process and injection molding. Results concluded an increasing value of tensile strength and modulus from 2.9 GPa to 3.6 GPa

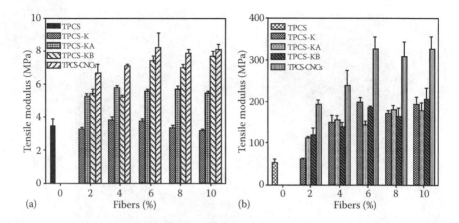

**FIGURE 11.15** Effect of fiber treatments on tensile strength (a) and tensile modulus (b). (From Siti Yasmine, Z.Z. et al., *Carbohydrate Polymers* 92: 2299–2305, 2013.)

and from 58 MPa to 71 MPa for nanocomposites with 5 wt% CNF. Meanwhile, the results show decreasing at maximum strain from 3.4% to 2.7% with additional CNF content. The mechanical properties of composites were shown in Table 11.5.

Cellulose nanocrystals showed great potential for unsaturated polyester nanocomposites, and were determined by Kargarzahed et al. (2015). To prepare cellulose nanocrystals, kenaf bast fibers underwent alkali and bleaching treatments in order to obtain cellulose, and were processed with acid hydrolysis. The aim of these studies was to determine the effects of CNC silane surface treatments on the mechanical properties, thermal behavior, and morphology of the nanocomposites. They concluded that tensile strength and modulus were much better compared to untreated CNC, even though it showed a decrease in tensile and modulus as the filler content increased, as observed in Figure 11.16.

**TABLE 11.5**
**Mechanical Properties of the PLA and Its Nanocomposites with Different Compositions**

| Materials | Tensile Modulus (GPa) | Tensile Strength (MPa) | Max Strain (%) |
|---|---|---|---|
| PLA | 2.9 ± 0.6 | 58.9 ± 0.5 | 3.4 ± 0.4 |
| PLA-CNF1 | 3.3 ± 0.4 | 63.1 ± 0.9 | 2.8 ± 0.3 |
| PLA-CNF2 | 3.4 ± 0.1 | 65.1 ± 0.6 | 2.7 ± 0.2 |
| PLA-CNF3 | 3.6 ± 0.7 | 71.2 ± 0.6 | 2.7 ± 0.1 |

# Development and Characterization of Kenaf Nanocomposites

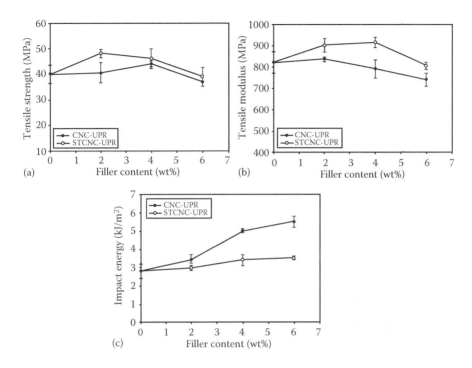

**FIGURE 11.16** Evolution of the (a) tensile strength, (b) tensile modulus, and (c) impact energy for the CNC-UPR and STCNC-UPR nanocomposites as a function of the filler content (CNC, cellulose nanocrystal; STCNC, silane-treated cellulose nanocrystal; UPR, unsaturated polyester resin). (From Kargarzadeh, H. et al., *Polymer* 56: 346–357, 2015.)

## REFERENCES

Abdul Khalil, H.P.S., Yusra, A.F.I., Bhat, A.H., & Jawaid M. 2010. Cell wall ultrastructure, anatomy, lignin distribution, and chemical composition of Malaysian cultivated kenaf fiber. *Ind. Crops Prod.* **31**: 113–121.

Abdul Khalil, H.P.S., Bhat, A.H., & Ireana Yusra A.F. 2012. Green composites from sustainable cellulose nanofibrils: A review. *Carbohydrate Polymers* **87**: 963–979.

Abe, K., Iwamoto, S., & Yano H. 2007. Obtaining cellulose nanofibers with a uniform width of 15 nm from wood. *Biomacromolecules* **8**(10): 3276–3278.

Abe, K., Nakatsubo, F., & Yano H. 2009. High-strength nanocomposite based on fibrillated chemi-thermomechanical pulp. *Composites Science and Technology* **69**(14): 2434–2437.

Abe, K. & Yano H. 2009. Comparison of the characteristics of cellulose microfibril aggregates of wood, rice straw and potato tuber. *Cellulose* **16**(6): 1017–1023.

Abe, K. & Yano H. 2010. Comparison of the characteristics of cellulose microfibril aggregates isolated from fiber and parenchyma cells of Moso bamboo (*Phyllostachyspubescens*). *Cellulose* **17**(2): 271–277.

Akil, H.M., Omar, M.F., Mazuki, A.A.M., Safiee, S., Ishak, Z.A.M., & Abu Bakar A. 2011. Kenaf fiber reinforced composites: A review. *Mater. Des.* **32**: 4107–4121.

Angles, M.N. & Dufresne A. 2001. Plasticized starch/tunic in whiskers nanocomposites materials. 2. Mechanical behavior. *Macromolecules* **34**(9): 2921–2931.

Araki, J., Wada, M., Kuga, S., & Okano T. 2000. Birefringent glassy phase of a cellulose microcrystal suspension. *Langmuir* **16**(6): 2413–2415.

Ashori, A., Raverty, W.D., & Harun J. 2006. Effect of totally chlorine free and elemental chlorine free sequences on whole stem kenaf (*Hibiscus cannabinus*) pulp characteristics. *Polym-Plast. Technol. Eng.* **45**(2): 205–211.

Ashori, A., Babaee, M., Jonoobi, M., & Hamzeh Y. 2014. Solvent-free acetylation of cellulose nanofibers for improving compatibility and dispersion. *Carbohydrate Polymers* **102**: 369–375.

Babaee, M., Jonoobi, M., Yahya, H., & Ashori A. 2015. Biodegradability and mechanical properties of reinforced starch nanocomposites using cellulose nanofibers. *Carbohydrate Polymers* **132**: 1–8.

Chakraborty, A., Sain, M., & Kortschot M. 2005. Cellulose microfibrils: A novel method of preparation using high shear refining and cryocrushing. *Holzforschung* **57**(1): 102–107.

Chakraborty, A., Sain, M., & Kortschot M. 2006. Reinforcing potential of wood pulp derived microfibres in a PVA matrix. *Holzforschung* **60**(1), 53–58.

Cheng, Q., Wang, S., & Han Q. 2010. Novel process for isolating fibrils from cellulose fibers by high-intensity ultrasonication. II. Fibril characterization. *Journal of Applied Polymer Science* **115**(5): 2756–2762.

Cheng, Q., Wang, S., Rials, T., & Lee S. 2007. Physical and mechanical properties of polyvinyl alcohol and polypropylene composite materials reinforced with fibril aggregates isolated from regenerated cellulose fibers. *Cellulose* **14**(6): 593–602.

Cheng, Q., Wang, S., & Rials T.G. 2009. Poly(vinyl alcohol) nanocomposites reinforced with cellulose fibrils isolated by high intensity ultrasonication. *Composites Part A: Applied Science and Manufacturing* **40**(2): 218–224.

Cheung, H.Y., Ho, M.P., Lau, K.T., Cardona, F., & Hui D. 2009. Natural fibre-reinforced composites for bioengineering and environmental engineering applications. *Composites Part* **40**: 655–663.

Cicala, G., Cristaldi, G., Recca, G., Ziegmann, G., El-Sabbagh, A., & Dickert M. 2009. Properties and performances of various hybrid glass/natural fibre composites for curved pipes. *Mater. Des.* **30**: 2538–2542.

Cunha, A. G. & Gandini A. 2010. Turning polysaccharides into hydrophobic materials: A critical review, Part 1. Cellulose. *Cellulose* **17**: 875–889.

Elazzouzi-Hafraoui, S., Nishiyama, Y., Putaux, J.L., Heux, L., Dubreuil, F., & Rochas C. 2007. The shape and size distribution of crystalline nanoparticles prepared by acid hydrolysis of native cellulose. *Biomacromolecules* **9**(1): 57–65.

Frenot, A., Henriksson, M.W., & Walkenstrom P. 2007. Electrospinning of cellulose based nanofibers. *Journal of Applied Polymer Science* **103**(3): 1473–1482.

Gousse, C., Chanzy, H., Cerrada, M.L., & Fleury E. 2004. Surface silylation of cellulose microfibrils: Preparation and rheological properties. *Polymer* **45**: 1569–1575.

Graupner, N., Herrmann, A.S., & Müssig J. 2009. Natural and man-made cellulose fibre reinforced polylactic acid (PLA) composites: An overview about mechanical characteristics and application areas. *Compos A. Appl. Sci. Manuf.* **40**: 810–821.

Hadi, A., Babak, G., Jalal, D., Ali, A.E., & Asghar K.A. 2015. Novel nanocomposites based on fatty acid modified cellulose nanofibers/poly(lactic acid): Morphological and physical properties. *Food Packaging and Shelf Life* **5**: 21–31.

Hanieh, K., Rasha, M.S., Ishak, A., Ibrahim, A., & Dufresne A. 2015. Toughened polyester cellulose nanocomposites: Effect of cellulose nanocrystals and liquid epoxidized natural rubber on morphology and mechanical properties. *Industrial Crops and Product* **72**: 125–132.

Hayashi, N., Kondo, T. & Ishihara M. 2005. Enzymatically produced nano-ordered short elements containing cellulose I crystalline domains. *Carbohydrate Polymers* **61**(2): 191–197.
Henriksson, M., Henriksson, G., Berglund, L.A. & Lindstrom T. 2007. An environmentally friendly method for enzyme-assisted preparation of microfibrillated cellulose (MFC) nanofibers. *European Polymer Journal* **43**: 3434–3441.
Hu, W., Chen, S., Xu, Q., & Wang H. 2011. Solvent-free acetylation of bacterial cellulose under moderate conditions. *Carbohydrate Polymers* **83**: 1575–1581.
Iwamoto, S., Kai, W., Isogai, T., Saito, T., Isogai, A., & Iwata T. 2010. Comparison study of TEMPO-analogous compounds on oxidation efficiency of woodcellulose for preparation of cellulose nanofibrils. *Polymer Degradation and Stability* **95**(8): 1394–1398.
Jawaid, M. & Abdul Khalil H.P.S. 2011. Cellulosic/synthetic fibre reinforced polymer hybrid composites: A review. *Carbohydr. Polym.* **86**: 1–18.
Jonoobi, M., Harun, J., Mathew, A.P., Hussein, M.Z.B., & Oksman K. 2010a. Preparation of cellulose nanofibers with hydrophobic surface characteristics. *Cellulose* **17**(2): 299–307.
Jonoobi, M., Harun, J., Mathew, A.P., & Oksman K. 2010b. Mechanical properties of cellulose nanofiber (CNF) reinforced polylatic acid (PLA) prepared by twin screw extrusion. *Composites Science and Technology* **70**: 1742–1747.
Jonoobi, M., Jalaludin, H., Alireza, S., Manjusri, M., & Kristiina O. 2009. Chemical composition, crystallinity, and thermal degradation of bleached and unbleached kenaf bast (*Hibiscus Cannabinus*) pulp and nanofibers. *Bioresources* **4**(2): 626–639.
Jonoobi, M., Jalaluddin, H., Paridah, M.T., Lukmanul, H.Z., Syeed, S., & Majid D.M. 2010c. Characteristic of nanofibers extracted from kenaf core. "Nanofibers from kenaf". *Bioresources* **5**(4): 2556–2566.
Jonoobi, M., Mathew, A.P., Abdi, M.M., Makinejad, M.D., & Oksman K. 2012. A comparison of modified and unmodified cellulose nanofibers reinforced poly-lactic acid (PLA) prepared by twin screw extrusion. *Journal of Polymers and the Environment* **20**(4): 991–997.
Kargarzedeh, H., Sheltami, R.M., Ishak, A., Ibrahim, A., & Dufresne A. 2015. Cellulose nanocrystal: A promising toughening agent for unsaturated polyester nanocomposite. *Polymer* **56**: 346–357.
Khalil, H.A., Ismail, H., Rozman, H., & Ahmad M. 2001. The effect of acetylation on interfacial shear strength between fibers and various matrices. *European Polymer Journal* **37**(5): 1037–1045.
Kim, C.W., Kim, D.S., Kang, S.Y., Marquez, M. & Joo Y.L. 2006. Structural studies of electrospun cellulose nanofibers. *Polymer* **47**(14): 5097–5107.
Klemm, D., Heublein, B., Fink, H.P., & Bohn A. 2005. *Angewandte Chemie International Edition* **44**: 3358.
Liu, W., Drzal, L.T., Mohanty, A.K., & Misra M. 2007. Influence of processing methods and fiber length on physical properties of kenaf fiber reinforced soy based biocomposite. Composites Part B: Engineering: Special Issue. *Biocomposites* **38**(3): 352–359.
Ma, Z., Kotaki, M., & Ramakrishna S. 2005. Electrospun cellulose nanofiber as affinity membrane. *Journal of Membrane Science* **265**(1–2): 115–123.
Matsumura, H., Sugiyama, J., & Glasser W.G. 2000. Cellulosic nanocomposites. I. thermally deformable cellulose hexanoates from heterogeneous reaction. *Journal of Applied Polymer Science* **78**: 2242–2253.
Mohd Edeerozey, A.M., Akil, M.H., Azhar, A.B., and Zainal Arifin, M.I. 2007. Chemical modification of kenaf fibers. *Materials Letters* **61**: 2023–2025.
Nakagaito, A.N. & Yano H. 2004. The effect of morphological changes from pulp fiber towards nano-scale fibrillated cellulose on the mechanical properties of high-strength plant fiber based composites. *Applied Physics A. Materials Science & Processing* **78**(4): 547–552.

Nakagaito, A.N. & Yano H. 2005. Novel high-strength biocomposites based on microfibrillated cellulose having nano-order-unit web-like network structure. *Applied Physics A: Materials Science & Processing* **80**(1): 155–159.

Nakagaito, A. & Yano H. 2008a. The effect of fiber content on the mechanical and thermal expansion properties of biocomposites based on microfibrillated cellulose. *Cellulose* **15**(4): 555–559.

Nakagaito, A. & Yano H. 2008b. Toughness enhancement of cellulose nanocomposites by alkali treatment of the reinforcing cellulose nanofibers. *Cellulose* **15**(2): 323–331.

Nogi, M., Iwamoto, S., Nakagaito, A.N. & Yano H. 2009. Optically transparent nanofiber paper. *Advanced Materials* **21**(16): 1595–1598.

Paakko, M., Ankerfors, M., Kosonen, H., Nykanen, A., Ahola, S. & Osterberg M. 2007. Enzymatic hydrolysis combined with mechanical shearing and high-pressure homogenization for nanoscale cellulose fibrils and strong gels. *Biomacromolecules* **8**(6): 1934–1941.

Pan, M., Zhou, X., & Chen M. 2013. Cellulose nanowhiskers isolation and properties from acid hydrolysis combined with high pressure homogenization. *BioResources* **8**(1).

Parikh, D.V., Calamari, T.A., & Sawhney A.P.S., Blanchard EJ, Screen FJ, Warnock M et al. 2002. Improved chemical retting of kenaf fibers. *Textile Res J.* **72**(7): 618–24.

Pettersson, L., Kvien, I., & Oksman K. 2007. Structure and thermal properties of poly (laticacid)/cellulose whickers nanocomposites materials. *Compost Science and Technology* **67**(11): 2535–2544.

Rassmann, S., Paskaramoorthy, R., & Reid R.G. 2010. Effect of resin system on the mechanical properties and water absorption of kenaf fibre reinforced laminates. *Mater. Des.* **32**: 1399–1406.

Ribot, N.M.H., Ahmad, Z., & Mustaffa N.K. 2011. Mechanical properties of kenaf fiber composite using co-cured in-line fiber joint. *Int. J. Eng. Sci. Technol.* **3**(4): 3526–3534.

Saba, N. Paridah, M.T., & Jawaid M. 2015. Mechanical properties of kenaf fiber reinforced polymer composite: A review.

Saheb, D.N., & Jog J.P. 1999. Natural fiber polymer composites: A review. *Adv. Polym. Technol.* **18**(4): 351–363.

Saito, T., Hirota, M., and Isogai, A. 2009. Individualization of nano-sized plant cellulose fibrils by direct surface carboxylation using TEMPO catalyst under neutral conditions. *Biomacromolecules* **10**(7): 1992–1996.

Saito, T., Kimura, S., Nishiyama, Y., & Isogai A. 2007. Cellulose nanofibers prepared by TEMPO-mediated oxidation of native cellulose. *Biomacromolecules* **8**(8): 2485–2491.

Saito, T., Nishiyama, Y., Putaux, J.L., Vignon, M., & Isogai A. 2006. Homogeneous suspensions of individualized microfibrils from TEMPO-catalyzed oxidation of native cellulose. *Biomacromolecules* **7**(6): 1687–1691.

Sassi, J.F. & Chanzy H. 1995. Ultrastructural aspects of the acetylation of cellulose. *Cellulose* **2**: 111–127.

Seng, Y.L., Ahmad, S.H., Rasid, R., Noum, S.Y., Hock, Y.C., & Tarawneh M.A. 2011. Effect of liquid natural rubber (LNR) on the mechanical properties of LNR toughened epoxy composite. *Sains Malaysiana* **40**: 679–683.

Siqueira, G., Bras, J., and Dufresne, A. 2010. Cellulosic bionanocomposites: A review of preparation, properties and application. *Polymers* **2**: 728–765.

Siró, I. & Plackett D. 2010. Microfibrillated cellulose and new nanocomposite materials: A review. *Cellulose* **17**(3): 459–494.

Siti Yasmine, Z.Z., Ishak, A., Hanieh, K., Ibrahim, A., & Dufresne A. 2013. Potential of using multiscale kenaf fibers as reinforcing filler in cassava starch-kenaf biocomposites. *Carbohydrate Polymers* **92**: 2299–2305.

Sreekala, M., Goda, K., & Devi P. 2008. Sorption characteristics of water, oil and diesel in cellulose nanofiber reinforced corn starch resin/remie fabric composites. *Composites Interfaces* **15**(2–3): 281–299.

Surip, S.N., Wan Jaafar, W.N.R., Azmi, N.N., & Anwar U.M.K. 2012. Microscopy observation on Nanocellulose from kenaf fiber. *Advanced Materials Research* **488–489**: 72–75.

Wang, S. & Cheng Q. 2009. A novel process to isolate fibrils from cellulose fibers by high intensity ultrasonication. Part 1. Process optimization. *Journal of Applied Polymer Science* **113**(2): 1270–1275.

Wang, B. & Sain M. 2007. Dispersion of soybean stock-based nanofiber in a plastic matrix. *Polymer International* **56**(4): 538–546.

Yousif, B.F., Shalwan, A., Chin, C.W., & Ming K.C. 2012. Flexural properties of treated and untreated kenaf/epoxy composites. *Mater. Des.* **40**: 378–385.

Zampaloni, M., Pourboghrat, F., Yankovich, S.A., Rodgers, B.N., Moore, J., & Drzal L.T. 2007. Kenaf natural fiber reinforced polypropylene composites: A discussion on manufacturing problems and solutions. *Composites Part A* **38**(6):1569–1580.

Zhao, H.P., Feng, X. Q., & Gao H. 2007. Ultrasonic technique for extracting nanofibers from nature materials. *Applied Physics Letters* **90**: 73–112.

# 12 Concurrent Design of Kenaf Composite Products

*M.R. Mansor and S.M. Sapuan*

## CONTENTS

| | | |
|---|---|---|
| 12.1 | Introduction | 205 |
| 12.2 | Concurrent Design of Kenaf Composite Products | 207 |
| | 12.2.1 Conceptual Design | 208 |
| | 12.2.2 Material Selection | 210 |
| | 12.2.3 Manufacturing Process Selection | 214 |
| 12.3 | Concurrent Design Case Study on Kenaf Composite Shelf Bracket | 216 |
| | 12.3.1 Problem Identification and Formulation of the General Problem | 217 |
| | 12.3.2 Formulation of General Solution | 218 |
| | 12.3.3 Formulation of Specific Solutions | 220 |
| | 12.3.4 Conceptual Design Development | 220 |
| 12.4 | Conclusion | 222 |
| References | | 223 |

## 12.1 INTRODUCTION

Today's growing concern for environmental sustainability has spurred great effort in terms of research and development by many industries to introduce new environmental friendly products into the market. Among the various strategies is material substitution, whereby recyclable and renewable materials replace conventional engineering materials which lack sustainability. Natural fiber composites (NFC) are some of the most applied materials from new product development applications, especially in the effort to replace synthetic-based composites. NFCs provide higher recyclability and biodegradability compared to synthetic composites due to the use of recyclable and renewable sources of reinforcement and matrix constituents (Razali et al. 2015). Moreover, the renewable raw materials also provide added advantages to new products especially in term of lower raw material cost (Yahaya et al. 2015). Furthermore, natural fibers, which make up the reinforcement constituents of the final composites, also enabled reduced manufacturing tool wear and very low negative health effects to the operators (such as skin and respiratory irritations), which finally resulted in lower overall product and operation cost (Araujo et al. 2014).

In general, the NFC market is divided into two major segments, which are wood fibers and non-wood fibers. Wood fibers are the most common fibers which are mostly utilized for building and construction applications, whereas non-wood fibers, such as kenaf, sisal, jute, and hemp, are mainly applied for automotive component production (Ray 2015). A statistical survey data in 2012 for the European market revealed that kenaf utilization in the automotive industry has contributed up to 8% or 6400 tons of the total 80,000 tons plastic composites used for automotive applications. This utilization is the third largest after flax (25%) and hemp (19%) composites (Biobased News 2013). The data showed high utilization of NFCs in automotive industry and the competiveness of the kenaf composites when compared to other commercially available NFCs for similar applications. It is noted that the use of NFCs can improve the component's lightweight properties due to the low density material properties of the natural fibers which is the focus of the automotive industry worldwide in the race towards producing new generation of lightweight and low fuel consumption vehicles. In addition, the potential of higher utilization is also wide open for other product applications such as for the building and constructions (producing fencing, panels, and frames) as well in sporting and electronics applications (producing rackets, snowboards, mobile phones, and laptop cases) (Lucintel 2011).

One of the key elements in producing NFC products is the product design and development process. The design and development of NFC products involves multidisciplinary activities which encompass defining customer requirements, creating the design, verifying the design engineering aspects such as strength and functionality, manufacturing of the product either as a prototype or in bulk, and selling the completed product to the intended user (Sapuan 2010). In this case, the overall process can be simplified and organized into an overall product design and development framework. One of the most applied product design and development framework is called Total Design (Pugh 1991). In this framework, all the activities related to product development are divided into six main categories, which are the order of market investigation, product design specifications, conceptual design, detail design, manufacturing, and finally sale of the product. The Total Design framework is created in sequential arrangement to reflect the process of creating the product from scratch until finish.

As the product design and development process evolved in order to cater to high customer expectations in many aspects such as cost and quality, the conventional sequential flow of creating the product became unfavorable and was gradually replaced with parallel product development flow termed as concurrent design. The main disadvantage of the sequential design method is that it cannot support design changes efficiently as the feedback between all product development processes are slowly communicated due to the step-by-step process. As a result, the changes made are not quickly visible to other linking sections, hence delaying the process of adapting the necessary changes to meet the requirements made by other stakeholders (such as within internal processes or between the external stakeholder and the internal processes). The concurrent design method in the other hand is based on cross linking the same project goal with different processes within the product development stages, thus enabling quicker information flow and design change adaptation between the respected stakeholders. In general, the concurrent design method is a systematic approach towards product design integration. It is also considered a customer-oriented design method in order to successfully meet the customer expectation on the end product (Hsiao 2002). The process flow is arranged in parallel so that all related activities can start simultaneously based on the same design descriptions by the customer. The concurrent design method utilized technological advances, such as computerized integration systems to manage the project activities and provide efficient communication between the cross-functional teams.

Based on the aforementioned description, the concurrent design method can be successfully adopted in NFC product development process. NFC product development is especially related to kenaf composites, and the product requires the simultaneous consideration of different aspects which are materials, design, and manufacturing in the early stage of the design process. Thus, the implementation of the concurrent design method in NFC product development can contribute in shorter design cycle and reducing the product design time. Hence, faster product introduction to the market can be achieved especially for new products, as well as increasing the overall product profit and competitiveness. In addition, the application of concurrent design methods can also reduce design error and late design changes in the product development stages, thus contributing to greater potential in reducing the risk and cost of the product (AGDA 2015). Furthermore, the quality and speed of decision making between the related stakeholders can also be improved in order to successfully produce products which meet the customer needs.

In Chapter 12, the implementation of concurrent design method for kenaf composites' product development is presented. The focus is given for the conceptual design stage of the product development process. Chapter 12 explains the concurrent design activities related to kenaf composites' product development which are conceptual design development and selection, material selection, and manufacturing process selection. Moreover, a case study on conceptual design development of kenaf composite products is also presented in the end of Chapter 12 to provide better understanding on the practical application of the concurrent design methods and how they can assist in creating innovative solutions to cater for the structural strength limitation of the kenaf composite materials.

## 12.2 CONCURRENT DESIGN OF KENAF COMPOSITE PRODUCTS

Chandrasegaran et al. (2013) stated that product design activities often deal with ill-defined objectives, whereby design specifications are normally refined as the product development stages progress. Moreover, they also emphasized that product development activities also involve complex and iterative processes, which need the systematic and efficient coordination between respective departments such as computerized systems in order to produce successful products which meet customer expectations. A similar situation also applied for NFC product design, particularly kenaf composite products, due to high amounts of data which need to be processed and incorporated into the design process, such as material properties, product functionality, target product costs, and sustainability performance.

Pickering et al. (2016) listed several factors contributing to the mechanical performance of NFC products such as fiber selection (which includes type of fiber, cultivation, and harvesting parameters, fiber aspect ratio and fiber content, fiber dispersion and orientation, matrix selection, composite manufacturing process and type of treatment used, composites interfacial strength, and the porosity of final composites. From a design perspective, two main elements are related based on Pickering et al.'s suggestions which are material selection and manufacturing process selection. In another report, Busch (2012) stated that design of NFC products requires different perspective compared to conventional engineering materials whereby factors such as good part/component design as well as suitable processing and materials selection (such as polymers and additives) plays an important role in the product development process.

Overall, there are generally three main elements involved in concurrent design related to conceptual design stage of NFC products which are concept design development and selection, material selection, and manufacturing process selection. The material selection can be further expanded into screening and ranking process. Moreover, both processes also cover two main components that makes up any natural fiber composite material; the reinforcement material (natural fibers) and the matrix material (resins). Further explanation on each of the conceptual design elements stated are explained in the next section.

### 12.2.1 Conceptual Design

The conceptual design process in the concurrent design method involves two main activities, which are the concept design development and concept design selection. The conceptual design process is concerned with the objective to determine the shape and form of the final composites product based on the given product design specifications (PDS) (Davoodi et al. 2011). The PDS is a technical document whereby all the design criteria are grouped together to serve as a reference document for the designer in developing the product. The PDS consists of information such as the intended product final performance, aesthetic appearance, desired product weight, etc. In the concurrent design method, the PDS elements are derived from customer requirements for the product.

There are many concurrent design tools which were developed and applied in order to capture the customer expectations into design requirements such as the quality function deployment (QFD) method (Vezzetti et al. 2016) and Kano model (Lin et al. 2011). Both tools use quantitative survey techniques to systematically determine the voice of the customer and translate it into the product design requirements. Using these methods, the product can be design to precisely in line with the customer needs whether in term of functionality, quality, cost, etc. As a result of embedding the customer expectations as the design requirements, the correct decision of the design solution which meet the requirements can be made as early as possible in the whole product development process, thus increasing the chances of product success when introduced into the market.

After determining the design guidelines and developing the product design specifications, the next process in conceptual design of composite products is design concept development. In general, this stage is performed to translate the design requirement into ideas in the geometrical shape of the product. There are several methods which can be applied to perform the task. One of the method is theory of inventive problem solving (TRIZ) which was developed G. Altshuller in the 1940s. The method is based on solving contradictions for the design problem to obtain the best design solution (Deimel 2011). The main advantages of the TRIZ method are due to its simple and very structured problem definition and solution development process (algorithm). Apart from that, the TRIZ method also enabled the process of analyzing engineering system evolution trends, which is very useful in assisting product designers to anticipate the product evolution over time, and to take necessary action during the design process to incorporate the trends into workable product functions (Yu & Fan 2012). The TRIZ method consists of five different approaches in solving the contradictions, which are classified according to the type of the problem model such as engineering contradiction and substance field models. Each problem model is assigned to a distinctive tool and solution model (such as 40 inventive principles and 76 specific standard inventive solutions) to solve the defined problem and obtain the desired solution. Table 12.1 shows the overview of the TRIZ model of problem and their respective tools and models of solutions applicable for the concept development process related to kenaf composite products.

There are many reported literatures on the application of the TRIZ method for concept development of various products. Yeh et al. (2011) implemented the TRIZ together with a four-phase QFD method to produce new conceptual designs of computer notebooks. At different QFD phases, new innovative technological solutions were able to be produced using the TRIZ methods such as energy efficient display, notebook base assembly design, circuit board design, and shop-floor preventive maintenance method. Yamashina et al. (2002) also reported the application of integrated the TRIZ and QFD towards the development of new washing machine design in order to improve operational aspects of the products such as washing mechanisms. On the other hand, Zhang et al. (2014) applied integrated QFD-TRIZ methods for producing a new conceptual design of kitchen stoves based on the voice of the customer, several innovative product functions which are focused on

**TABLE 12.1**
**Summary of the TRIZ Approach in Problem Identification and Idea Generation**

| Problem Model | Tool | Solution Model |
|---|---|---|
| Engineering contradiction | Contradiction matrix | 40 inventive principles |
| Physical contradiction | Separation, satisfaction, bypass | 40 inventive principles |
| Function model | Scientific effect | Specific scientific effects |
| Substance field model | System of standard inventive solution | 76 standard inventive solutions |
| Algorithm of inventive problem solving (ARIZ) | All of the above | All of the above |

*Source:* San et al., *TRIZ: Systematic Innovation in Manufacturing.* Selangor: Firstfruits Sdn. Bhd. 2009.

improving the ergonomic aspects of the product which were able to be generated using the TRIZ method. Similar integrated applications of QFD and TRIZ were also shown as successful in assisting product designers to come out with technological solutions for medical applications. Melgoza et al. (2012) implemented the combination of several concurrent design methods such as attribute listing, QDF, and the TRIZ to develop a new tracheal stent device concept design in term of new geometrical configuration and materials based on the requirements gathered from physicians, hospital administrators, and patients. They also further manufactured the newly develop stent design using an additive manufacturing process.

Despite many successful reports on the application of concurrent design methods for conceptual design purpose (especially the TRIZ), very few mentioned on its application for NFC product development. Among the available resources about the application of the TRIZ for the conceptual design process of NFC products was by Mansor et al. (2014) on the conceptual design of automotive products. In their report, the new conceptual design of an automotive parking brake lever using hybrid kenaf/glass fiber composites was developed using the integration of the TRIZ and morphological chart methods. The TRIZ engineering contradiction matrix and 40 inventive principle tools were applied for problem identification and the idea generation process, while the morphological chart was employed to further assist in concept design generation based on functionality of the product. They also further applied the analytic hierarchy process (AHP) method to select the best design concept for the new hybrid NFC automotive parking brake lever. In the other report, Mansor et al. (2015) also implemented a similar integrated the TRIZ morphological chart method for the process of generating a new conceptual design of automotive spoiler products made from kenaf composites. Using a combination of kenaf composites at the outer layer of the component and a polymer based sandwich between the layers, the new composite's spoiler solution was able to be produced for improved lightweight performance while maintaining component structural strength. In addition, Sapuan and Mansor (2015) also demonstrated the application of the TRIZ method for conceptual design development of new laptop casings made from NFCs. Apart from improving the product weight, cost, and recyclability feature based on NFC advantages, other innovative design solutions were also generated for the NFC products using the TRIZ method, such as foldable covers to improve portability and a built-in cooling fan on the product for self-service cooling performance and component integration. The aforementioned examples showed the capability of the concurrent design method in assisting product designers dealing with NFC, especially kenaf composites, in order to break the psychological inertia and stigma of low NFC inherent structural properties to produce innovative solution to address the limitation for their product design in the conceptual design stage of the product.

### 12.2.2 Material Selection

Material selection is of one the major tasks in the early stage of the product development process. Material selection is often involved complex decision making due to multiple requirements or criteria that need to be satisfied by the selected materials to ensure all aspects of the product can performed successfully as per design intents.

# Concurrent Design of Kenaf Composite Products 211

In addition, the difficulty in achieving the best decision for the best selected materials also arises from the pool of potential materials to be selected, whereby all of the listed candidate materials available are qualified in performing the required tasks. For instance, Ramesh (2016) listed 49 type of natural fibers (including kenaf fibers) suitable for the production of NFC. Deciding on the best type of natural fiber for a specific application based on a wide variety of candidate materials requires a comprehensive supporting tool to analyze all the data simultaneously. The supporting tool must also be able to include information on different attributes, such as material properties for the individual type of fibers.

In addition, decision making in material selection also needs to satisfy various requirements from various stakeholders simultaneously, which conflict with each other most of the time, as shown in Figure 12.1. Handling with this sort of challenges requires a systematic and scientific decision-making approach in material selection in order to obtain the best decision for the intended solution while complying with all the conflicting requirements.

In general, decision-making in kenaf composite material selection involves two stages, namely the screening stage and ranking stage. In most cases, the material selection process starts with a screening stage where potential materials that able to perform the requirements are selected from all possible materials categories based on the product design requirements (metals, ceramics, polymers, etc.). The screening process can be performed using several methods, such as a pass-fail method or the Ashby chart method. The goal of the screening process is to eliminate the non-compliable materials based on the design requirements and shortlist the applicable materials. The screening process involved identification of the minimum, maximum,

**FIGURE 12.1** Conflicts arising from improper material selection. (From Sapuan, S.M., Mansor, M.R., *Materials & Design*, 58, 161–167, 2014.)

or range of values of the design requirements that need to be complied by the candidate materials. For example, the chosen materials must have a minimum tensile strength of 100 MPa, have a raw material price of maximum RM 100 per kg, and be able to operate at range of temperature between 15°C to 50°C. By screening the materials based on the given requirements, the final suitable material that passes the selection test is then selected as the suitable material to be used to produce the product.

Among the reported literature on the application of screening methods for NFC application was a comparative study of 12 pineapple leaf fiber varieties for use as mechanical reinforcement in polymer composites (Neto et al. 2015). In their report, the screening process was conducted using two fiber parameters, which are fiber strength and onset oxidation temperature for each of the pineapple leaf fiber cultivars investigated. Elsewhere, Shah (2014) reported the application of comprehensive Ashby chart for materials screening selection of plant fiber composites. The use of the Ashby chart was shown able to provide quick and informative screening results to designers in deciding on the most appropriate candidate materials based on their tensile property, cost, weight, and eco-impact performance.

However, in several cases, product designers may not need to perform the decision-making and screening process because they have already identified the list of potential materials suitable for the intended application. Then, the next challenge which arises is in deciding the best material to be selected among available candidate materials. This stage in decision making is categorized as the ranking or choosing stage, whereby using a particular method, all of the listed candidate materials are analyzed based on their degree of capability in complying with the selection requirements. The ranking stage can be further elaborated in two sub-stages, which are first identifying the degree of importance of each selection criteria with respect to the selection goal and later rank the candidate materials based on the highest performance to the selection criteria according to their inherent material properties. The material which scores the highest point and rank first among the other candidate materials is finally chosen as the best material to be applied for the intended application.

There are many ranking methods that have been developed and applied in material selection. The AHP method developed by Saaty has been applied in the material selection of automotive bumper beam design for automotive parking brake lever design (Hambali et al. 2010) as well as the material selection of building materials for sustainable construction (Akadiri et al. 2013). The preferences by similarity to the ideal solution (TOPSIS) method developed by Hwang and Yoon were applied in the raw material selection for pulping and papermaking (Anupam et al. 2014); material selection of grinding wheel abrasive under fuzzy approach (Maity and Chakraborty 2013); and material selection of piezoelectric material for energy harvester, sensor, and actuator applications (Chauhan and Vaish 2013). Similarly, the selection process was also performed using *Vlsekriterijumska Optimizacija I Kompromisno Resenje* (VIKOR) method developed by Opricovic for the material selection of automotive interior instrument panels (Girubha and Vinodh 2012) as well as the material selection of metallic biomaterials for femoral component construction in knee-joint implants (Jahan and Edwards 2013). Apart from that, the preference ranking organization method for enrichment evaluation (PROMETHEE) method developed by

Brans and Vincke was also applied for material selection purposes of journal bearing materials (Peng and Xiao 2013) and the selection of tool holder materials for hard milling operation (Çalişkan et al. 2013). Another applied material selection method is the *Elimination Et Choix Traduisant la Realité* (in English is translated as "Elimination and Choice Expressing the Reality") or ELECTRE method developed by Benayoun, Roy, and Sussman. Among the reported materials selection project utilizing the ELECTRE method were for the selection of the best material for thermal heat conductor components (Shanian and Savadogo 2006a) and bipolar plates materials selection for polymer electrolyte fuel cells (Shanian and Savadogo 2006b).

Presently, it is observed that there is a growing trend in the study on the material selection of NFCs in general, including kenaf composites for various product applications. Among the available literature of the NFC material selection by Sapuan et al. (2011) are 29 NFC candidate materials for automotive dashboard panel applications. In their report, the AHP method was applied to select the best type of NFC material, whereby the final results revealed that kenaf–polypropylene composites was the best candidate material for dashboard panel construction, followed by flax–epoxy composites.

In another report, Mansor et al. (2013) also demonstrated that the application of multi-criteria decision making method in performing materials selection process for developing automotive parking brake lever component. Their study involved analysis on the selection of the best type of natural fibers to be hybridized with glass reinforced polypropylene composites for the component construction. The AHP method was applied in the decision-making process and the final results showed that kenaf fibers are the most suitable candidate material to form the hybrid composites for automotive parking brake lever components.

Moreover, the application of the AHP method as the selection tool for hybrid natural–aramid fiber composite material selection was also reported by Yahaya et al. (2014). What is unique is their initiative to introduce hybrid natural fiber composites into military applications, specifically towards the construction of spall liner components which involved ballistic impact loading conditions. Their findings showed that kenaf fiber is the best candidate material for the military spall liner component construction compared to other natural fiber alternatives involved in the selection process.

Furthermore, Ahmed Ali et al. (2015) reported the development of expert decision-making systems based on the AHP method to assist product designers in executing the material selection process involving NFCs for automotive interior component applications. Their new expect decision system strived on improving the environmental performance and sustainability aspects in the decision-making process. It was shown from their report that using the expert decision system, hemp-reinforced polypropylene composites emerged as the best candidate material for automotive interior components based on the product design specifications.

Apart from that, the material selection process involving NFCs for automotive friction material application was demonstrated by Mustafa et al. (2014). In their report, a weighted decision matrix method was applied in the material selection process to select the best type natural fibers to be used for the new eco-friendly friction material formulation. The end results from the analysis showed that kenaf fibers scored the highest among other natural fiber candidate materials for the automotive friction material applications.

In addition, Al-Oqla et al. (2016) introduced a new decision-making model on NFC material selection for automotive applications. Their new model was developed to systematically aid the selection process involving nonwoven natural fiber-reinforced polypropylene composites. A total of 15 candidate materials, which consisted of various types of natural fibers, were reinforced with a polypropylene matrix and were analyzed using the new decision-making model. The final results showed that date palm–polypropylene composites is the most suitable candidate material for the intended application.

Based on the aforementioned literatures, it can be observed that the material selection process is more focused to the composite reinforcement constituents for product development applications. One of the reasons is that the fibers act as the main load-bearing component for the overall composite material, henceforth the success of developing structural products relies heavily on the appropriate type of fibers to be used to endure the structural loads exerted to the product. However, the overall NFC product performance capability is also contributed to the effectiveness of the matrix material in holding together the fibers and distributing the applied load evenly to the whole structure.

With regards to the importance of matrix material for the overall NFC product performance, it was also observed that there several reports which involved material selection on NFC matrix material. Mansor et al. (2014) applied an integrated AHP–TOPSIS method for the material selection of thermoplastic matrix material towards the construction of hybrid composites for automotive parking brake levers. An AHP method was applied to determine the weight of the selection criteria whereas the TOPSIS method was used to rank the thermoplastic material candidates included in the selection process. Moreover, Mansor et al. (2014) also applied TOPSIS multi-criteria decision-making method for thermoset matrix material selection. The material selection was performed to select the best thermoset matrix for automotive bumper beam applications.

### 12.2.3 Manufacturing Process Selection

The third element which involved in the concurrent design of kenaf composite products is the manufacturing process selection. The manufacturing process is crucial in order to produce the required final shape of the product with good quality and a competitive cost. There are many criteria involved in selection of appropriate manufacturing process for kenaf and other natural fiber composite products such as the final composites' shape, complexity, size, production volume, production cycle, tooling cost, labor intensity, and composite raw material processing characteristics (such as sheet molding compound, yarn, pre-pregs, and mat) (Ho et al. 2012). In addition, the manufacturing process selection of kenaf composite products also requires considerations on the desired composites volume contents, porosity, and reinforcement form (in term of final fiber length and fiber orientation within the product). Most importantly, the selection of the appropriate process depends on the type of matrix for the composites, either using a thermoplastic or thermosetting matrix (Shah 2013).

# Concurrent Design of Kenaf Composite Products

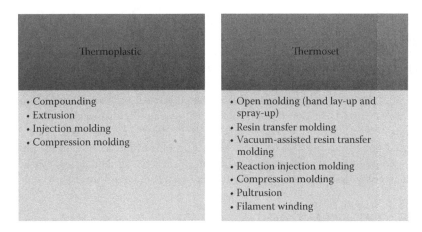

**FIGURE 12.2** Classification of natural fiber composite manufacturing processes. (From Composites World, Fabrication Methods, 2014. Retrieved February 15, 2015 from http://www.compositesworld.com/articles/fabrication-methods.)

Figure 12.2 classified several major manufacturing processes related to natural fiber composites according to the type of matrix used. The list can be further divided into open molding and closed molding processes. Open molding process such as hand lay-up and semi-automated spray-up are traditional composite manufacturing process. In the other hand, closed molding processes such as extrusion, compression molding, injection molding, and resin transfer molding are a new class of modern and automated composite manufacturing processes which are currently among the most applied processes to produce natural fiber composite products (Faruk et al. 2014).

For an open molding process, the advantages are cheap raw materials, unskilled labor, and tooling costs, which is suitable to produce low performance and large components. However, the disadvantages of the open molding process is a high void (porosity) occurrence and low fiber volume fraction in the final product (Summerscales and Grove 2014). Injection molding and compression molding are the most applied manufacturing processes for natural fiber composites due to high volume production which in return able to provide lower product cost. Injection molding also enables a fast production cycle, but affects the final fiber length and distribution (Liu et al. 2007). Moreover, compression molding is also preferred in manufacturing kenaf composite products because of process simplicity, but results in challenges to control the final fiber distribution during processing (Akil et al. 2011). As for pultrusion and filament winding processes, they enable long continuous production to be obtained as well as fast and good final product quality. In addition, the pultrusion process also enables constant cross sectional part geometry with a high fiber volume content (Fairuz et al. 2015). Closed molding processes (such as resin transfer moulding, vacuum assisted resin transfer molding, and reaction injection molding) are also favorable for medium-sized production which also offers low porosity performance of the final product (Shah 2013).

Currently, the application of concurrent design tools in the performing manufacturing process selection for composite products is still limited. One notable effort was reported on the manufacturing process selection for composite automotive bumper beam products using the AHP method (Hambali et al. 2009). In their report, the selection criteria used for the decision-making process during the conceptual design stage of the product was production characteristics, cost, product design geometry, type of raw material, ease of maintenance, and availability of the labor and equipment. Based on their analysis, the injection molding process emerged as the most preferred composite manufacturing process to produce the bumper beam product compared to other manufacturing candidates, such as structural resin injection molding, compression molding, and the resin transfer molding process.

## 12.3 CONCURRENT DESIGN CASE STUDY ON KENAF COMPOSITE SHELF BRACKET

To demonstrate the application of concurrent design on kenaf composite products, a case study was performed on the development of new interior shelf bracket components using kenaf composites. In general, interior shelf brackets are household components used as the supporting structure to hold wall mounted shelves. The component is usually made from steel due to the material's high structural strength and reliability which are crucial to hold the shelves firmly at the mounted location. The structural component geometry is produced in L-shape form due to the mounting design requirement between the mounted shelves and the wall.

Despite the acceptable current performance of steel-based shelf brackets, there is room for potential improvement which may be further applied to the component itself in order to provide additional advantages. Improvements can be made in terms of reducing the component's weight by replacing the steel with a lower density material to produce shelf brackets. A lighter shelf bracket can help to reduce the overall component load, which subsequently requires a smaller screw size to be used to mount the bracket to the wall. Apart from that, replacing steel with a cheaper material may also further reduce the component cost, due to the fact steel prices are rising in the current market due to depleting availability of iron to produce the steel.

Thus, based on the aforementioned situations, natural fiber composites, especially kenaf composites, may become an excellent candidate material to substitute the use of steel in producing shelf bracket components. Kenaf composites have a lower density compared to steel, and are therefore able to produce lighter shelf bracket components. Furthermore, kenaf fibers are a cheap and renewable source of raw material which able to grant lower cost to produce the same component compared to steel (which is a non-renewable raw material). The use of reinforced plastic in product design such as kenaf composites also provides additional advantages in terms of product design, whereby more geometric design configurations can be easily manufactured using plastic manufacturing processes compared to steel manufacturing processes.

In spite of the kenaf composites, many advantages such as a light weight, low cost, and design flexibility for the intended application, one notable design limitation of the natural fiber composites are their inherent lower mechanical strength. As the result, kenaf composite applications are limited to non-structural products, whereby low load bearing requirements are present. The conflicting situation between low structural strength versus a light weight and low cost now appeared in the process of applying kenaf composites to make shelf brackets. There are many potential solutions which have already been developed to address the low structural strength limitation of natural fiber composites, such as through chemical treatments (using fiber treatment agents and coupling agents) and hybridization techniques (combining natural fibers with synthetic fibers in a single matrix) to enhance the natural fiber composites' mechanical properties. In this case study, the performance limitation of kenaf composites is dealt with the application of concurrent design methods which proposed an innovative solution for the product with minimal changes to the material itself. The concurrent design method which will be explained involved the application of TRIZ and morphological chart methods for problem root cause identification and idea generation.

### 12.3.1 Problem Identification and Formulation of the General Problem

In the first stage of the conceptual design process, the determination on the root cause of the problem is conducted using the TRIZ method. Based on the problem description explained in the previous section, it is noted that the problem which need to be solve in order to apply kenaf composites for shelf bracket products is low mechanical strength to handle the existing bending load from the shelves and other products placed on top of it. Low mechanical strength will result in structural failure in term of bending the kenaf composite shelf bracket. In addition, low mechanical strength material properties also induce a lower reliability to the product, which may cause the product to fail when introduced to the market. In return, if the problems of low strength and low reliability are solved, the new kenaf composite shelf bracket will be lighter and cheaper compared to steel shelf bracket.

In order kickstart the idea generation process to solve the problem, the TRIZ recommended that the problem be defined in a general statement. The purpose is to eliminate any mental barrier or obstacle in the creative thinking process due to the use of technical jargon of the designer. The TRIZ general problem statement included information on improving aspects of the new design, the negative aspects of the design, and the manipulative variable related to the product. In this case study, the improved aspect of utilizing kenaf composites for the new product was lower product weight, whereas the resulting worsening aspect of utilizing kenaf composites for the same product was causing low mechanical strength and reliability. Thus, based on the TRIZ recommendation, the general problem statement for the case is formulated as below:

"If the shelves bracket uses kenaf composites, then the weight of the shelves bracket is reduced, but strength and reliability of the shelves bracket are also reduced."

Based on the above general statement, the improving parameter which needed to be obtained was reducing the weight of the product, and at the same time also caused worsening conditions in terms of the product's strength and reliability. In normal problem-solving practices, the contradiction is dealt with trade-off, whereby only one aspect can be satisfied in the same time and scarifying the other aspect. However, using the TRIZ, both conflicting aspects are solved in the same time, eliminating the need for trade-off. In this case, the TRIZ helped the designer come up with the solution for the new kenaf composites shelf bracket which will be lighter, have a good structural strength, and perform reliably, which is required for safe and functional product operation.

### 12.3.2 Formulation of General Solution

In the second stage of the TRIZ method, the problem model identified previously is matched with the TRIZ 39 engineering parameters which will later be used to construct the TRIZ contradiction matrix. Based on the contradiction matrix, general solution to solve the problem (or contradiction) is determined based on TRIZ 40 Inventive Principles. The TRIZ 39 engineering parameters and 40 Inventive Principles are powerful tools created based on analysis of patents solutions. The improving and worsening parameters identified in the previous general problem statement are matched with the TRIZ 39 engineering parameters. In this case, the improving parameter for lighter shelf brackets according to TRIZ is the weight of stationary object (parameter no. 2 in the list of TRIZ 39 engineering parameter), whereas the resulting worsening parameter for reduce of product strength and reliability are represented by strength (parameter no. 14 in the list of TRIZ 39 engineering parameter) and reliability (parameter no. 27 in the list of TRIZ 39 engineering parameter), respectively.

After using the new TRIZ terminology for presenting the improving and worsening parameters related to the problem, another TRIZ tool, the contradiction matrix, is later applied. The contradiction matrix is a 39 × 39 matrix which listed all the 39 engineering parameters according to the improving and worsening conditions for the problem. The designer then only needs to select the appropriate improving and worsening parameters as arranged in the contradiction matrix, and identify the recommended TRIZ 40 inventive solution principles at the location where the two improving and worsening parameters intersect. Table 12.2 shows the outcome of this exercise and the resulting general solutions recommended using the TRIZ 40 inventive principles to solve the kenaf composites shelf bracket problem.

As shown in Table 12.2, for each contradiction statement, there are four TRIZ inventive principles which are recommended to be applied to solve the problem. For example, in order to reduce the weight of the product while maintaining adequate product strength, the recommended inventive solution is mechanical substitution, taking out, preliminary action, and cheap short-living objects. Further examination of the description of each general solution principle revealed that there are several recommendations that can be applied, such as taking out. Taking out can be further interpreted as removing or eliminating something from the current design. Thus, at the end of the exercise, potential solutions to solve the problem were obtained using the TRIZ tool. The process towards acquiring general solutions using the TRIZ

## TABLE 12.2
## TRIZ Contradiction Matrix Corresponding 40 Inventive Principles

| Improving Parameters | Worsening Parameters | Recommended Inventive Principles to Be Used and Their Descriptions |
|---|---|---|
| No. 2: Weight of stationary object | No. 14: Strength | No. 28: Mechanics substitution<br>• Replace a mechanical means with a sensory (optical, acoustic, taste, or smell) means<br>• Use electric, magnetic, and electromagnetic fields to interact with the object<br>• Change from static to movable fields, from unstructured fields to those having structure<br>• Use fields in conjunction with field-activated (e.g., ferromagnetic) particles<br>No. 2: Taking out<br>• Separate an interfering part or property from an object, or single out the only necessary part (or property) of an object.<br>No. 10: Preliminary action<br>• Perform, before it is needed, the required change of an object (either fully or partially).<br>• Prearrange objects such that they can come into action from the most convenient place and without losing time for their delivery<br>No. 27: Cheap short living objects<br>• Replace an inexpensive object with a multiple of inexpensive objects, comprising certain qualities (such as service life, for instance) |
| No. 2: Weight of stationary object | No. 27: Reliability | No. 10: Preliminary action<br>• -same as above-<br>No. 28: Mechanics substitution<br>• -same as above-<br>No. 8: Anti-weight<br>• To compensate for the weight of an object, merge it with other objects that provide lift<br>• To compensate for the weight of an object, make it interact with the environment (e.g., use aerodynamic, hydrodynamic, buoyancy and other forces)<br>No. 3: Local quality<br>• Change an object's structure from uniform to non-uniform, change an external environment (or external influence) from uniform to non-uniform<br>• Make each part of an object function in conditions most suitable for its operation<br>• Make each part of an object fulfil a different and useful function |

*Source:* San et al., *TRIZ: Systematic Innovation in Manufacturing.* Selangor: Firstfruits Sdn. Bhd. 2009.

method is described using simple and systematic algorithms which helped designers source for faster solutions compared to conventional brainstorming methods.

### 12.3.3 FORMULATION OF SPECIFIC SOLUTIONS

As explained in the previous section, the TRIZ method has enable solutions to be identified using a systematic and guided method. However, it should also be noted that the TRIZ method only proposed general solutions to the problem with a very abstract definition of the meaning of each inventive principle. Hence, it is up to the designer to further interpret the general solution into a more specific solution which can directly be applied to the problem. In this stage, the knowledge and experience of the designer plays an important role in the success of translating the general solution into specific solution. For instance, the general solution of taking out can be the guide to removing some portion of the product to reduce its weight without jeopardizing its structural strength. The specific recommendation that can be developed is to apply pocket feature on the shelf bracket body. Another example is by taking general solution's local quality, whereby specific solutions that can be generated for the design are by using nonuniform bracket thickness (a thicker cross section at a high stress concentration area) and adding ribs to provided additional reinforcement to the initial structure.

Further effectiveness on developing the specific solution based on the TRIZ general inventive principles can also be added by the aid of another concurrent design tool which is the morphological chart method. In general, the morphological chart is a visual representation of potential solution and the solution is organized into a chart for ease of use. Using the morphological chart, more detailed ideas regarding the initial specific solution can be generated. For example, adding ribs to the kenaf composite shelf bracket can be further expanded into two categories of rib shapes, either straight or curve. Thus, now for the same specific solution, there are two more sub-solutions that can be applied for similar problems. Expanding the list of potential solutions in more detail can help designers to thoroughly look for new ideas and promoting more creative thinking to solve the problem. Figure 12.3 shows the morphological chart developed for this case study based on the general inventive solutions obtained using the TRIZ method.

### 12.3.4 CONCEPTUAL DESIGN DEVELOPMENT

The expansion of new ideas for the kenaf composite shelf bracket products can be clearly observed through the morphological chart in order to retain the improving features of reducing the weight while in the same time maintaining the structural strength and product reliability. For each of the inventive principles, there are at least two design solutions that can be applied to represent it, such as for the case of local quality and taking out. Based on the morphological chart developed as shown in Figure 12.3, the final conceptual design for the new kenaf composite shelf bracket can be easily generated by combining each of the individual features to make up the final product. For example, new conceptual design of the product can be developed with integration of straight rib, uniform cross section, and no pocket features. Another new

# Concurrent Design of Kenaf Composite Products

| General solution (TRIZ 40 inventive principles) | Specific solution (by the designer) | Solution (Design feature) | | |
|---|---|---|---|---|
| | | 1 | 2 | 3 |
| No. 3: Local quality | To introduce rib at the center of the L-shape bracket to provide reinforcement against bending load | No rib  | Straight rib  | Curve rib  |
| | To change the bracket and cross-section from uniform thickness to non-uniform thickness (the joint location of the L-shape is made thicker from the rest of the part) | Uniform cross section  | Non-uniform cross section  | |
| No. 2: Taking out | To introduce hollow pockets at the bracket to reduce the weight of the bracket | No pocket (solid) 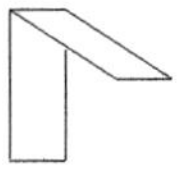 | Uniform pocket 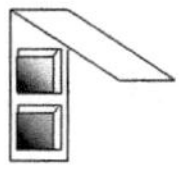 | Non-uniform pocket  |

**FIGURE 12.3** Morphological chart method for kenaf composite shelf bracket design.

conceptual design for similar product can also be developed with the combination of other features as listed in the morphological chart such as straight rib, nonuniform cross section, and pocket design. Thus, through the same morphological chart as shown in Figure 12.3, numerically it is possible to produce up to 18 new conceptual designs for a similar product just by combining the individual design features. Figure 12.4 shows several examples of new conceptual designs feasible to produce the kenaf composites shelf brackets using different combinations of design features listed in the morphological chart.

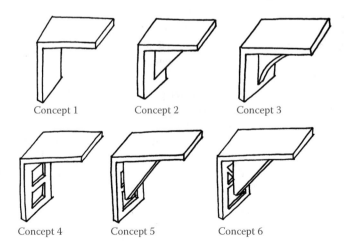

**FIGURE 12.4** Several new conceptual designs of the kenaf composite shelf brackets.

A higher number of conceptual designs created will provide more solution options to solve similar problems. As a result, the designer can conduct an in-depth analysis of the solution options and select the best solution which will be able to successfully fulfill the intended design requirements. To aid the selection process, the designer can further apply multi-criteria decision-making method (such as AHP) to perform decision-making on the best conceptual design solution. It should also be noted, more solutions in term of conceptual designs can also be generated by expanding the solution options for each of the design features in the morphological chart. For instance, expanding the proposed three-rib design solution option to include tang rib design cross rib design will generate a total of five specific solution options for similar inventive principle. This is easily possible using group-based brainstorming method compared to individual brainstorming method. Involvement of designers with higher level of experience and know-how will also accelerate the brainstorming process and expand the morphological chart more effectively.

## 12.4 CONCLUSION

Based on previous explanations, it can be concluded that the task of designing NFCs products, especially kenaf composite products, requires a different perspective by product designers whereby multiple elements need to be considered simultaneously with the involvement of many stakeholders. Variation of attributes and alternatives for each element add up to the complexity and challenges to product design in designing successful composite products. For that reason, a concurrent design method was able perform the required task in designing kenaf composite products and deliver the expected outcome based on systematic and scientific methods. The overall process of concurrent design of kenaf composite products involved three main elements, which are conceptual design development, material selection, and manufacturing process selection. Furthermore, customer requirements also need to be incorporated into the concurrent design process as early as in the conceptual design stage of the product. Among the tools available within the concurrent design umbrella for designers to choose from in performing the required tasks are the TRIZ for conceptual design development and MCDM methods for decision making purposes. Besides that, the case study presented in this chapter also amplified the practicality and usefulness of the concurrent design methods in solving the design barrier created by the inherent mechanical properties of kenaf composite materials toward improving product performance, especially related to higher load bearing applications. Finally, by following systematic concurrent design process, products made from kenaf composites can be successfully developed to fully comply with the intended design specifications. Moreover, the concurrent design methods also helped designers achieve the intended design goal for the kenaf composite products at greater speed, better outcome quality, and lower development cost.

# REFERENCES

AGDA. (2015). Concurrent Design. Retrieved February 15, 2016, from http://www.adga.ca/en/services/9-concurrent-design.html.

Ahmed Ali, B. A., Sapuan, S. M., Zainudin, E. S., & Othman, M. (2015). Implementation of the expert decision system for environmental assessment in composite materials selection for automotive components. *Journal of Cleaner Production*, 107(0), 557–567.

Akadiri, P. O., Olomolaiye, P. O., & Chinyio, E. A. (2013). Multi-criteria evaluation model for the selection of sustainable materials for building projects. *Automation in Construction*, 30(0), 113–125.

Akil, H. M., Omar, M. F., Mazuki, A. A. M., Safiee, S., Ishak, Z. A. M., & Abu Bakar, A. (2011). Kenaf fiber reinforced composites: A review. *Materials & Design*, 32(8–9), 4107–4121.

Al-Oqla, F. M., Sapuan, S. M., Ishak, M. R., & Nuraini, A. A. (2016). A decision-making model for selecting the most appropriate natural fiber–polypropylene-based composites for automotive applications. *Journal of Composite Materials*, 50(4), 543–556.

Anupam, K., Lal, P. S., Bist, V., Sharma, A. K., & Swaroop, V. (2014). Raw material selection for pulping and papermaking using TOPSIS multiple criteria decision making design. *Environmental Progress and Sustainable Energy*, 33(3), 1034–1041.

Araujo, M. A. M. D., Neto, S., Hage, E., Mattoso, L. H. C., & Marconcini, J. M. (2014). Curaua leaf fiber (Ananas comosus var. erectifolius) reinforcing poly (lactic acid) biocomposites: Formulation and performance. *Polymer Composites*, 36(8), 1–11.

Biobased News. (2013). Biocomposites: 350,000 t production of wood and natural fiber composites in the European Union in 2012. Retrieved February 15, 2016, from http://bio-based.eu/news/biocomposites/.

Busch, J. (2012). Plastics go on natural-fiber diet. Retrieved February 15, 2015 from http://machinedesign.com/archive/plastics-go-natural-fiber-diet.

Çalişkan, H., Kurşuncu, B., Kurbanoğlu, C., & Güven, Ş. Y. (2013). Material selection for the tool holder working under hard milling conditions using different multi criteria decision making methods. *Materials and Design*, 45, 473–479.

Chandrasegaran, S. K., Ramani, K., Sriram, R. D., Horváth, I., Bernard, A., Harik, R. F., & Gao, W. (2013). The evolution, challenges, and future of knowledge representation in product design systems. *Computer-Aided Design*, 45(2), 204–228.

Chauhan, A. & Vaish, R. (2013). Material selection for piezoelectric devices. *Advanced Science, Engineering and Medicine*, 5(7), 715–719.

Composites World. (2014). Fabrication Methods. Retrived February 15, 2015 from www.compositesworld.com/articles/fabrication-methods.

Davoodi, M. M., Sapuan, S. M., Ahmad, D., Aidy, A., Khalina, A., & Jonoobi, M. (2011). Concept selection of car bumper beam with developed hybrid bio-composite material. *Materials & Design*, 32(10), 4857–4865.

Deimel, M. (2011). Relationships between TRIZ and classical design methodology. *Procedia Engineering*, 9(0), 512–527.

Fairuz, A. M., Sapuan, S. M., Zainudin, E. S., & Jaafar, C. N. A. (2015). Pultrusion Process of Natural Fibre-Reinforced Polymer Composites. In S. M. Sapuan, M. Jawaid, Y. Nukman, & M. E. Hoque (Eds.), *Manufacturing of Natural Fibre Reinforced Polymer Composites* (pp. 217–231). Switzerland: Springer.

Faruk, O., Bledzki, A. K., Fink, H.-P., & Sain, M. (2014). Progress report on natural fiber reinforced composites. *Macromolecular Materials and Engineering*, 299(1), 9–26.

Girubha, R. J. & Vinodh, S. (2012). Application of fuzzy VIKOR and environmental impact analysis for material selection of an automotive component. *Materials and Design*, 37, 478–486.

Hambali, A., Sapuan, S. M., Ismail, N., & Nukman, Y. (2009). Composite manufacturing process selection using Analytical Hierarchy Process. *International Journal of Mechanical and Materials Engineering*, 4(1), 49–61.

Hambali, A., Sapuan, S. M., Ismail, N., & Nukman, Y. (2010). Material selection of polymeric composite automotive bumper beam using analytical hierarchy process. *Journal of Central South University of Technology*, 17(2), 244–256.

Ho, M., Wang, H., Lee, J.-H., Ho, C., Lau, K., Leng, J., & Hui, D. (2012). Critical factors on manufacturing processes of natural fibre composites. *Composites Part B: Engineering*, 43(8), 3549–3562.

Hsiao, S. (2002). Concurrent design method for developing a new product. *International Journal of Industrial Ergonomics*, 29(1), 41–55.

Jahan, A. & Edwards, K. L. (2013). VIKOR method for material selection problems with interval numbers and target-based criteria. *Materials & Design*, 47, 759–765.

Lin, C.-S., Chen, L.-S., & Hsu, C.-C. (2011). An innovative approach for RFID product functions development. *Expert Systems with Applications*, 38(12), 15523–15533.

Liu, W., Drzal, L., Mohanty, A., & Misra, M. (2007). Influence of processing methods and fiber length on physical properties of kenaf fiber reinforced soy based biocomposites. *Composites Part A: Applied Science and Manufacturing*, 38, 352–359.

Lucintel. (2011). Opportunities in Natural Fiber Composites. Retrieved from http://www.lucintel.com/OpportunitiesinNaturalFiberComposites.pdf.

Maity, S. R. & Chakraborty, S. (2013). Grinding wheel abrasive material selection using fuzzy TOPSIS method. *Materials and Manufacturing Processes*, 28(4), 408–417.

Mansor, M. R., Sapuan, S. M., Hambali, A., Zainudin, E. S., & Nuraini, A. A. (2014). Materials Selection of Hybrid Bio-Composites Thermoset Matrix for Automotive Bumper Beam Application using Topsis Method. *Advances in Environmental Biology*, 8(8), 3138–3442.

Mansor, M. R., Sapuan, S. M., Hambali, A., Zainudin, E. S., & Nuraini, A. A. (2015). Conceptual design of kenaf polymer composites automotive spoiler using TRIZ and morphology chart methods. *Applied Mechanics and Materials*, 761(0), 63–67.

Mansor, M. R., Sapuan, S. M., Zainudin, E. S., Nuraini, A. A., & Hambali, A. (2013). Hybrid natural and glass fibers reinforced polymer composites material selection using Analytical Hierarchy Process for automotive brake lever design. *Materials & Design*, 51, 484–492.

Mansor, M. R., Sapuan, S. M., Zainudin, E. S., Nuraini, A. A., & Hambali, A. (2014). Application of integrated AHP-TOPSIS method in hybrid natural fiber composites materials selection for automotive parking brake lever component. *Australian Journal of Basic and Applied Sciences*, 8(5), 431–439.

Mansor, M. R., Sapuan, S. M., Zainudin, E. S., Nuraini, A. A., & Hambali, A. (2014). Conceptual design of kenaf fiber polymer composite automotive parking brake lever using integrated TRIZ–Morphological Chart–Analytic Hierarchy Process method. *Materials & Design*, 54(0), 473–482.

Melgoza, E. L., Serenó, L., Rosell, A., & Ciurana, J. (2012). An integrated parameterized tool for designing a customized tracheal stent. *Computer-Aided Design*, 44(12), 1173–1181.

Mustafa, A., Abdollah, M. F., Shuhimi, F. F., Ismail, N., Amiruddin, H., & Umehara, N. (2014). Selection and verification of kenaf fibres as an alternative friction material using Weighted Decision Matrix method. *Materials & Design*, 67(0), 577–582.

Neto, A. R. S., Araujo, M. A. M., Barboza, R. M. P., Fonseca, A. S., Tonoli, G. H. D., Souza, F. V. D., ... Marconcini, J. M. (2015). Comparative study of 12 pineapple leaf fiber varieties for use as mechanical reinforcement in polymer composites. *Industrial Crops and Products*, 64, 68–78.

Peng, A. H. & Xiao, X. M. (2013). Material selection using PROMETHEE combined with analytic network process under hybrid environment. *Materials and Design*, 47, 643–652.

Pickering, K. L., Efendy, M. G. A., & Le, T. M. (2016). A review of recent developments in natural fibre composites and their mechanical performance. *Composites Part A: Applied Science and Manufacturing*, 83, 98–112.

Pugh, S. (1991). *Total Design: Integrated Methods for Successful Product Engineering*. Wokingham, England: Addison-Wesley Publishing.

Ramesh, M. (2016). Kenaf (*Hibiscus cannabinus* L.) fibre based bio-materials: A review on processing and properties. *Progress in Materials Science*, 78(0), 1–92.

Ray, D. (2015). State-of-the-Art Applications of Natural Fiber Composites in the Industry. In R. D. S. G. Campilho (Ed.), *Natural Fiber Composites* (pp. 319–340). Boca Raton: CRC Press.

Razali, N., Salit, M. S., Jawaid, M., Ishak, M. R., & Lazim, Y. (2015). A Study on Chemical Composition, Physical, Tensile, Morphological, and Thermal Properties of Roselle Fibre: Effect of Fibre Maturity. *Bioresources*, 10(1), 1803–1823.

San, Y. T., Jin, Y. T., & Li, S. C. (2009). *TRIZ: Systematic Innovation in Manufacturing*. Selangor: Firstfruits Sdn Bhd.

Sapuan, S. M. (2010). *Concurrent Engineering for Composites*. Serdang: UPM Press.

Sapuan, S. M., Kho, J. Y., Zainudin, E. S., Leman, Z., Ahmed Ali, B. A., & Hambali, A. (2011). Materials selection for natural fiber reinforced polymer composites using analytical hierarchy process. *Indian Journal of Engineering & Materials Sciences*, 18(4), 255–267.

Sapuan, S. M. & Mansor, M. R. (2014). Concurrent engineering approach in the development of composite products: A review. *Materials & Design*, 58, 161–167.

Sapuan, S. M. & Mansor, M. R. (2015). Design of Natural Fiber-Reinforced Composite Structures. In *Natural Fiber Composites* (pp. 255–278). Boca Raton: CRC Press.

Shah, D. U. (2013). Developing plant fibre composites for structural applications by optimising composite parameters: A critical review. *Journal of Materials Science*, 48(18), 6083–6107.

Shah, D. U. (2014). Natural fibre composites: Comprehensive Ashby-type materials selection charts. *Materials and Design*, 62, 21–31.

Shanian, A. & Savadogo, O. (2006a). A material selection model based on the concept of multiple attribute decision making. *Materials & Design*, 27(4), 329–337.

Shanian, A. & Savadogo, O. (2006b). ELECTRE I decision support model for material selection of bipolar plates for polymer electrolyte fuel cells applications. *Journal of New Materials for Electrochemical Systems*, 199, 191–199.

Summerscales, J. & Grove, S. (2014). Manufacturing methods for natural fibre composites. In A. Hodzic & R. Shanks (Eds.), *Natural Fibre Composites: Materials, Processes and Applications* (pp. 176–215). Cambridge; UK: Woodhead Publishing.

Vezzetti, E., Marcolin, F., & Guerra, A. L. (2016). QFD 3D: A new C-shaped matrix diagram quality approach. *International Journal of Quality & Reliability Management*, 33(2), 178–196.

Yahaya, R., Sapuan, S. M., Jawaid, M., Leman, Z., & Zainudin, E. S. (2015). Effect of layering sequence and chemical treatment on the mechanical properties of woven kenaf–aramid hybrid laminated composites. *Materials & Design*, 67(0), 173–179.

Yahaya, R., Sapuan, S. M., Leman, Z., & Zainudin, E. S. (2014). Selection of natural fibre for hybrid laminated composites vehicle spall liners using Analytical Hierarchy Process (AHP). *Applied Mechanics and Materials*, 564(0), 400–405.

Yamashina, H., Ito, T., & Kawada, H. (2002). Innovative product development process by integrating QFD and TRIZ. *International Journal of Production Research*, 40(5), 1031–1050.

Yeh, C. H., Huang, J. C. Y., & Yu, C. K. (2011). Integration of four-phase QFD and TRIZ in product R&D: A notebook case study. *Research in Engineering Design*, 22(3), 125–141.

Yu, H., & Fan, D. (2012). Man-made boards technology trends based on TRIZ evolution theory. *Physics Procedia*, 33(0), 221–227.

Zhang, F., Yang, M., & Liu, W. (2014). Using integrated quality function deployment and theory of innovation problem solving approach for ergonomic product design. *Computers & Industrial Engineering*, 76(0), 60–74.

# Index

## A

Acetylation, 188
Adhesion characteristics of kenaf fibers, 37–60
   adhesion principle, 38–41
   adhesion properties of kenaf, 43
   adsorption theory, 39
   bonding limitation in kenaf fibers, 53–54
   bonding mechanisms in natural fibers, 41–42
   buffering capacity, 47–49
   buffering capacity of kenaf, 50–53
   capillary effects, 47
   case hardening, 55
   chemical bonding, 40–41
   chemisorption theory, 39
   compatibility of kenaf fibers and polymers, 55
   diffusion theory, 40
   fiber morphology of kenaf, 42–43
   fiber-reinforced composites, 42
   future challenges, 55–58
   heat, 55
   hydrophilic surfaces, 46
   hydrophobic surfaces, 46
   kenaf fibers, 42
   lotus effect, 46
   mechanical interlocking, 38–39
   medium density boards, 42
   surface energy and contact angle, 44–46
   surface inactivation, 54–55
   wettability of kenaf, 49–50
   wetting and wettability, 46–47
Adsorption theory, 39
Alkaline–peroxide bleaching process, 189, 190
Alkali treatment, 70
Analytic hierarchy process (AHP), 210
Anthraquinone (AQ), 189
Armored vehicles (AVs), 161
Automotive components (interior), *see* Polypropylene reinforced with kenaf core fiber, effect of silica aerogel on (for interior automotive components)

## B

Ballistic properties, *see* Hybrid kenaf composites, ballistic properties of
Betel nut husk (BNH) fiber, 13
Biopolymers, 130, 178–179
Bleaching process, 174, 190
Buffering capacity, 47–49

## C

Capillary effects, 47
Carbon fiber–kenaf mat (CFKM), 123
Carboxymethyl cellulose (CMC), 174
Case hardening, 55
Catastrophic failure mode indicator (CFMI), 125–126
Cellulose-based composites from kenaf fibers, 169–184
   biopolymers, 178–179
   carboxymethyl cellulose, 174
   composites, 180–181
   electron beam irradiation treatment, 174
   extraction of cellulose from kenaf fibers, 173–175
   extraction of kenaf fibers, 170–173
   kenaf fibers, 169–170
   kenaf-reinforced epoxy, 181
   microcrystalline cellulose, 174
   thermoplastic composites, 175–176
   thermoset composites, 176–177
Cellulose nanocrystals (CNCs), 196
Cellulose nanofibers (CNFs), 187
CFE, *see* Crush force efficiency
CFMI, *see* Catastrophic failure mode indicator
Chemical bonding, 40–41
Chemisorption theory, 39
CMC, *see* Carboxymethyl cellulose
CNCs, *see* Cellulose nanocrystals
CNFs, *see* Cellulose nanofibers
Compression molding, 71, 215
Concurrent design of kenaf composite products, 205–226
   analytic hierarchy process, 210
   case study on kenaf composite shelf bracket, 216–222
   compression molding, 215
   conceptual design, 208–210
   conceptual design development, 220–222
   contradictions, 209, 219
   formulation of general solution, 218–220
   formulation of specific solutions, 220
   injection molding, 215
   manufacturing process selection, 214–216
   material selection, 210–214
   problem identification and formulation of the general problem, 217–218
   product design specifications, 208
   quality function deployment, 208

**227**

ranking methods, 212
theory of inventive problem solving, 209
Total Design, 206
Crashworthiness parameters, *see* Failure modes crashworthiness parameters of kenaf composite hexagonal tubes, effects of material types on
Crush force efficiency (CFE), 124

## D

Decortication, 24, 173
Derivative of thermogram (DTG) curves, 14
Diffusion theory, 40
Dynamic mechanical analysis (DMA), 72–73

## E

EBI treatment, *see* Electron beam irradiation treatment
Eco-friendly kenaf hybrid materials, 129–143
   biopolymers, 130
   characterizations, 132–133
   experiment, 131–133
   FTIR analysis of LDPE/TPS/KCF/HC composites after exposed to outdoor natural weathering, 138, 140
   halloysite clays, 130
   materials, 131
   morphological of LDPE/TPS/KCF/HC composites after exposed to outdoor natural weathering, 139–140
   outdoor weathering test, 132
   physical appearance of LDPE/TPS/KCF/HC composites after exposed to natural weathering, 133
   results and discussion, 133–141
   sago starch, 130
   sample fabrication, 131–132
   tensile properties of LDPE/TPS/KCF/HC composites after exposed to outdoor natural weathering, 133–137
E-glass fiber yarn and kenaf yarn (GFKY), 120
Electron beam irradiation (EBI) treatment, 174
Ethylene-propylene-diene monomer (EPDM), 100
Extrusion, 72

## F

Failure modes crashworthiness parameters of kenaf composite hexagonal tubes, effects of material types on, 113–128
   catastrophic failure mode indicator (CFMI), 125–126
   crush testing and crashworthiness parameters, 118
   effect of hybrid material in initial peak load, 124
   fabrication process, 117
   geometry and mandrel design, 115
   glass–carbon–glass/epoxy segmentation, 114
   load-displacement curves and failure mechanisms, 118–122
   results of CFKM, 118–119
   results of CFKY, 121
   results of crashworthiness, 122–126
   results of GFKM, 121–122
   results of GFKY, 120–121
   results of KYKM, 119
Fiber classification, 146
Fiber morphology of kenaf, 42–43
Fiber-reinforced composites (FRC), 42
Flexural strength, 106
Fourier Transform Infrared (FTIR), 74, 108
Fragment simulating projectiles (FSP), 152

## G

GFKY, *see* E-glass fiber yarn and kenaf yarn
Glass–carbon–glass/epoxy (GCG) segmentation, 114
Green building, 97
Greenhouse gas emissions, 1
Green technology, demand for, 186

## H

Halloysite clays (HCs), 130
Hemicellulose, 94
High density polyethylene (HDPE), 38
Hybrid composites, 98
Hybrid kenaf composites, ballistic properties of, 145–167
   armored vehicles, 161
   ballistic properties of kenaf–aramid hybrid composites, 154–157
   chemical treatment, 159–160
   classification of fibers, 146
   delamination, 155
   fabric architectures, 158–159
   factors affecting ballistic properties of kenaf hybrid composites, 157–160
   fiber content, 157–158
   fiber orientation, 157
   fragment simulating projectiles, 152
   hybridization, 146
   interlayer delamination, 155
   kenaf fiber hybrid composites, 150
   kenaf fiber-reinforced composites, 148–152
   kenaf–synthetic fiber hybrid composites, 150–152

# Index

measurement of ballistic properties of kenaf-aramid hybrid composites, 152–157
potential applications of kenaf-aramid hybrid composites in ballistics, 160–161
spall liners, 160
synthetic fibers, 146
temperature and loading rate, 158
Hydrophilic surfaces, 45, 188
Hydrophobic surfaces, 45, 188

## I

Injection molding, 72, 215

## K

Kenaf-aramid hybrid composites, *see* Hybrid kenaf composites, ballistic properties of
Kenaf bast fiber bundles (KBFB), 158
Kenaf-derived cellulose-reinforced (KDC) PLA composites, 75
Kenaf fiber structure and properties, 23–35
    anatomical properties, 24–28
    chemical properties, 28–30
    decortication, 24
    fiber bundles, 23
    kenaf stems, 24
    lignocellulosic fibers, 26
    mechanical properties, 30–32
    physical properties, 30
    sclerenchyma, 23
    wood plastic composite, 24
Kenaf polypropylene polymer composite (KPPC), 88
Kenaf-reinforced epoxy (KFRE), 181

## L

LDPE/TPS/KCF/HC composites, *see* Eco-friendly kenaf hybrid materials
Life cycle assessments (LCAs), 4–8
Lignin, 94
Lignocellulosic fibers, 26
Liquefied kenaf core (LKC), 100
Liquid epoxidized natural rubber (LENR), 196
Lotus effect, 46
Low density polyethylene (LDPE), 38, 131

## M

Mechanical interlocking, 38–39
Medium density boards (MDF), 42
Melamine urea formaldehyde (MUF), 47
Microcrystalline cellulose (MCC), 174
Moisture content criterion (MCC), 9
Montmorillonite nanoclay (MMT), 152

## N

Nanocomposites (kenaf), development and characterization of, 185–203
    acetylation, 188
    alkaline–peroxide bleaching process, 189, 190
    cellulose nanocrystals, 196
    cellulose nanofibers, 187
    extraction of cellulose nanofiber, 188–192
    extraction of nanocellulose from kenaf fibers, 188–194
    kenaf fibers, 185–187
    liquid epoxidized natural rubber, 196
    miscellaneous extraction methods of nanofibrils, 192–194
    nanofiber reinforced with composites, 194–199
    reinforcing agents, 185
    soda–anthraquinone pulping, 189, 190
    unsaturated polyester resin, 196
Natural fiber composites (NFCs), 1–22, 205
    advantages and disadvantages of, 8
    applications, 15–18
    betel nut husk fiber, 13
    compatibility of fibers and polymers, 9–11
    derivative of thermogram curves, 14
    development of, major issues in, 7–15
    factors influencing the composite performance, 14–15
    fillers, 16
    future developments, 18
    greenhouse gas emissions, 1
    life cycle assessments, 4–8
    moisture content criterion, 9
    natural fibers, 3–4
    sustainability, 1
    thermal stability of natural fibers, 11–14
    water absorption characteristics, 7–9
NFCs, *see* Natural fiber composites
Nitrogen oxide (NOx) emissions, 7

## P

PDS, *see* Product design specifications
Pectin, 95
Phenol formaldehyde (PF), 47
Pineapple leaf fibers (PALF), 101
Polyether ether ketone (PEEK), 38
Polylactic acid (PLA), 38, 75, 152, 158
Polypropylene (PP) reinforced with kenaf core fiber, effect of silica aerogel on (for interior automotive components), 81–92
    kenaf board fibricating, 85–86
    kenaf core processing, 85

kenaf polypropylene polymer composite, 88
   materials, 85
   mechanical properties, 86, 87–90
   methods, 85–86
   results and discussion, 87–90
   silica aerogel, 83
Polyvinyl alcohol (PVOH), 73
Product design specifications (PDS), 208; *see also* Concurrent design of kenaf composite products

## Q

Quality function deployment (QFD), 208

## R

Raw kenaf fiber UP composites (RK-UP-C), 105
Rolled homogenous armour (RHA), 161

## S

Sago starch, 73, 130
Scanning electron microscopy (SEM), 106–107, 171, 191
Sclerenchyma, 23
SEA, *see* Specific energy absorption
Silane treatment, *see* Unsaturated polyester composites (kenaf fiber), impact of silane treatment on properties of
Silica aerogel, *see* Polypropylene reinforced with kenaf core fiber, effect of silica aerogel on (for interior automotive components)
Soda–anthraquinone pulping, 189, 190
Spall liners, 160
Specific energy absorption (SEA), 123
Synthetic fibers, 6, 23, 61, 146

## T

Tensile strength, 105
Theory of inventive problem solving (TRIZ), 209
Thermogravimetric analysis (TGA), 11, 75
Thermoplastic composites, kenaf fiber-reinforced, 61–79
   alkali treatment, 70
   characterization of, 72–75
   compression molding, 71
   critical role effecting performance of natural fiber and kenaf fiber-reinforced polymer composites, 69–70
   dynamic mechanical analysis, 72–73
   extrusion, 72
   injection molding, 72
   kenaf fiber, 62–63
   kenaf fiber-reinforced thermoplastic polymers, 63–68
   low density polypropylene composites, 73
   polylactic acid composites, 75
   polypropylene composites, 72–73
   polyurethane composites, 73–75
   processing method, 71–72
Thermoplastic polyurethane (TPU), 74, 158
Thermoplastic sago starch (TPSS), 73
Thermoplastic starch (TPS) nanocomposites, 194
Total Design, 206
TRIZ, *see* Theory of inventive problem solving

## U

Ultra-high molecular weight polyethylene (UHMWPE) fibers, 161
Unsaturated polyester composites (kenaf fiber), impact of silane treatment on properties of, 93–111
   advantages of kenaf, 99–100
   advantages of natural fibers in composites, 98
   applications of natural fibers, 98
   benefits of natural fiber-reinforced composites, 96
   biocomposites, 97–98
   chemical treatment, 102–103
   drawback of kenaf to its composites, 102
   drawbacks of natural fiber-reinforced composites, 96–97
   ethylene-propylene-diene monomer, 100
   experimental results for kenaf composites, 103–107
   fiber composites, 95–96
   fibers, 98–102
   flexural strength, 106
   Fourier transform infrared spectroscopy analysis, 103–105
   green building, 97
   hemicellulose, 94
   hybrid composites, 98
   kenaf, 98–99
   lignin, 94
   liquefied kenaf core, 100
   mechanical properties, 105–106
   natural fiber composites, 95–98
   overview of composites, 95
   pectin, 95
   pineapple leaf fibers, 101
   properties of kenaf fibers, 99
   raw kenaf fiber UP composites, 105

# Index

scanning electron microscopy, 106–107
silane treatment, 102–103
specimen fabrication, 103
tensile strength, 105
trend in kenaf composites, 100–102
trends of fiber reinforced, 97
Unsaturated polyester resin (UPR), 196
Urea formaldehyde (UF), 47

## V

van der Waals forces, 39, 40, 46

## W

Wettability of kenaf, 49–50
Wood plastic composite (WPC), 24

PGMO 06/22/2018